James Watt

By

Andrew Carnegie

James Watt

CHAPTER I

CHILDHOOD AND YOUTH

James Watt, born in Greenock, January 19, 1736, had the advantage, so highly prized in Scotland, of being of good kith and kin. He had indeed come from a good nest. His great-grandfather, a stern Covenanter, was killed at Bridge of Dee, September 12, 1644, in one of the battles which Graham of Culverhouse fought against the Scotch. He was a farmer in Aberdeen shire, and upon his death the family was driven out of its homestead and forced to leave the district.

Watt's grandfather, Thomas Watt, was born in 1642, and found his way to Crawford's Dyke, then adjoining, and now part of, Greenock, where he founded a school of mathematics, and taught this branch, and also that of navigation, to the fishermen and seamen of the locality. That he succeeded in this field in so little and poor a community is no small tribute to his powers. He was a man of decided ability and great natural shrewdness, and very soon began to climb, as such men do. The landlord of the district appointed him his Baron Bailie, an office which then had important judicial functions. He rose to high position in thetown, being Bailie and Elder, and was highly respected and honored. He subsequently purchased a home in Greenock and settled there, becoming one of its first citizens. Before his death he had established a considerable business in odds and ends, such as repairing and provisioning ships; repairing instruments of navigation, compasses, quadrants, etc., always receiving special attention at his hands.

The sturdy son of a sturdy Covenanter, he refused to take the test in favor of prelacy (1683), and was therefore proclaimed to be "a disorderly school-master officiating contrary to law." He continued to teach, however, and a few years later the Kirk Session of Greenock, notwithstanding his contumacy, found him "blameless in life and conversation," and appointed him an Elder, which required him to overlook not only religious observances, but the manners and morals of the people. One of the most important of these duties was to provide for the education of the young, in pursuance of that invaluable injunction of John Knox, "that no father, of what estate or condition that ever he may be, use his children at his own fantasies, especially in their youth hood, *but all must be compelled to bring up their children in learning and virtue.*" Here we have, at its very birth, the doctrine of compulsory education for all the people, the secret of Scotland's progress. Great as was the service Knox rendered in the field ecclesiastical, probably what he did for the cause of public education excels it. The man who proclaimed that he would never rest until there was a public school in every parish in Scotland must stand for all time as one of the foremost of her benefactors; probably, in the extent and quality of the influence he exerted upon the national character through universal compulsory education, the foremost of all.

The very year after Parliament passed the Act of 1696, which at last fulfilled Knox's aspirations, and during the Eldership of Watt's grandfather, Greenock made prompt provision for her parish school, in which we may be sure the old "teacher of mathematics" did not fail to take a prominent part.

Thomas Watt's son, the father of the great inventor, followed in his father's footsteps, after his father's death, as shipwright, contractor, provider, etc., becoming famous for his skill in the making of the most delicate instruments. He built shops at the back of his house, and such were the demands upon him that he was able to keep a number of men, sometimes as many as fourteen, constantly at work. Like his father, he became a man of position and influence in the community, and was universally esteemed. Prosperity attended him until after the birth of his famous son. The loss of a valuable ship, succeeded by other misfortunes, swept away most of the considerable sum which he had made, and it was resolved that James would have to be taught a trade, instead of succeeding to the business, as had been the intention.

Fortunate it was for our subject, and especially so for the world, that he was thus favored by falling heir to the best heritage of all, as Mr. Morley calls it in his address to the Midland Institute—"the necessity at an early age to go forth into the world and work for the means needed for his own support." President Garfield's verdict was to the same effect, "The best heritage to which a man can be born is poverty." The writer's knowledge of the usual effect of the heritage of milliondom upon the sons of millionaires leads him fully to concur with these high authorities, and to believe that it is neither to the rich nor to the noble that human society has to look for its preservation and improvement, but to those who, like Watt, have to labor that they may live, and thus make a proper return for what they receive, as working bees, not drones, in the social hive. Not from palace or castle, but from the cottage have come, or can come, the needed leaders of our race, under whose guidance it is to ascend.

We have a fine record in the three generations of the Watts, great-grandfather, grandfather and father, all able and successful men, whose careers were marked by steady progress, growing in usefulness to their fellows; men of unblemished character, kind and considerate, winning the confidence and affection of their neighbors, and leaving behind them records unstained.

So much for the male branch of the family tree, but this is only half. What of that of the grandmothers and mothers of the line—equally important? For what a Scotch boy born to labor is to become, and how, cannot be forecast until we know what his mother is, who is to him nurse, servant, governess, teacher and saint, all in one. We must look to the Watt women as carefully as to the men; and these fortunately we find all that can be desired. His mother was Agnes Muirhead, a descendant of the Muirheads of Lachop, who date away back before the reign of King David, 1122. Scott, in his "Minstrelsy of the Scottish Border," gives us the old ballad of "The Laird of Muirhead," who played a great part in these unsettled days.

The good judgment which characterised the Watts for three generations is nowhere more clearly shown than in the Lady James Watt's father courted and finally succeeded in securing for his wife. She is described as a gentlewoman of reserved and quiet deportment, "esteemed by her neighbours for graces of person as well as of mind and heart, and not less distinguished for her sound sense and good manners than for her cheerful temper and excellent housewifery." Her likeness is thus drawn, and all that we have read elsewhere concerning her confirms the truth of the portrait. Williamson says that the lady to whom he (Thomas Watt) was early united in marriage was Miss Agnes Muirhead, a gentlewoman of good understanding and superior endowments, whose excellent management in household affairs would seem to have contributed much to the order of her establishment, as well as to the every-day happiness of a cheerful home. She is described as having been a person above common in many respects, of a fine womanly presence, ladylike in appearance, affecting in domestic arrangements—according to our traditions—what, it would seem was considered for the time, rather a superior style of living. What such a style consisted in, the reader shall have the means of judging for himself. One of the author's informants on such points more than twenty years ago, a venerable lady, then in her eighty-fifth year, was wont to speak of the worthy Bailie's wife with much characteristic interest and animation. As illustrative of what has just been remarked of the internal economy of the family, the old lady related an occasion on which she had spent an evening, when a girl, at Mrs. Watt's house, and remembered expressing with much *naïveté* to her mother, on returning home, her childish surprise that "Mrs. Watt had *two* candles lighted on the table!" Among these and other reminiscences of her youth, one venerable informant described James Watt's mother, in her eloquent and expressive Doric, as, "a braw, braw, woman—none now to be seen like her."

There is another account from a neighbor, who also refers to Mrs. Watt as being somewhat of the grand lady, but always so kind, so sweet, so helpful to all her neighbors.

The Watt family for generations steadily improved and developed. A great step upward was made the day Agnes Muirhead was captured. We are liable to forget how little of the original strain of an old family remains in after days. We glance over the record of the Cecils, for instance, to find that the present Marquis has less than one four-thousandth part of the Cecil blood; a dozen marriages have each reduced it one-half, and the recent restoration of the family to its pristine greatness in the person of the late Prime Minister, and in his son, the brilliant young Parliamentarian, of whom great things are predicted already, is to be credited equally to the recent infusion into the Cecil family of the entirely new blood of two successive brides, daughters of commoners who made their own way in the world. One was the mother of the late statesman, the other his wife and the mother of his sons. So with the Watt family, of which we have records of three marriages. Our Watt, therefore, had but one-eighth of the original Watt strain; seven-eighths being that of the three ladies who married into the family. Upon the entrance of a gentlewoman of Agnes Muirhead's qualities hung important results, for she was a remarkable character with the indefinable air of distinction, was well educated, had a very wise head, a very kind heart and all the sensibility and enthusiasm of the Celt, easily touched to fine issues. She was a Scot of the Scots and a storehouse of border lore, as became a daughter of her house, Muirhead of Lachop.

Here, then, we have existing in the quiet village of Greenock in 1736, unknown of men, all the favorable conditions, the ideal soil, from which might be expected to appear such "variation of species" as contained that rarest of elements, the divine spark we call genius. In due time the "variation" made its appearance, now known as Watt, the creator of the most potent instrument of mechanical force known to man.

The fond mother having lost several of her children born previously was intensely solicitous in her care of James, who was so delicate that regular attendance at school was impossible. The greater part of his school years he was confined most of the time to his room. This threw him during most of his early years into his mother's company and tender care. Happy chance! What teacher, what companionship, to compare with that of such a mother! She taught him to read most of what he then knew, and, we may be sure, fed him on the poetry and romance upon which she herself had fed, and for which he became noted in after life. He was rated as a backward scholar at school, and his education was considered very much neglected.

Let it not be thought, however, that the lad was not being educated in some very important departments. The young mind was absorbing, though its acquisitions did not count in the school records. Much is revealed of his musings and inward development in the account of a visit which he paid to his grandmother Muirhead in Glasgow, when it was thought that a change would benefit the delicate boy. We read with pleasant surprise that he had to be sent for, at the request of the family, and taken home. He kept the household so stirred up with his stories, recitations and continual ebullitions, which so fairly entranced his Grannie and Grandpa and the cousins, that the whole household economy was disordered. They lost their sleep, for "Jamie" held them spellbound night after night with his wonderful performances. The shy and contemplative youngster who had tramped among the hills, reciting the stirring ballads of the border, had found an admiring tho astonished audience at last, and had let loose upon them.

To the circle at home he was naturally shy and reserved, but to his Grannie, Grandpa, and Cousins, free from parental restraint, he could freely deliver his soul. His mind was stored with the legends of his country, its romance and poetry, and, strong Covenanters as were the Watts for generations, tales of the Martyrs were not wanting. The heather was on fire within Jamie's breast. But where got you all that *perferidum Scotorum*, my wee mannie—that store of precious nutriment that is to become part of yourself and remain in the core of your being to the end, hallowing and elevating your life with ever-increasing power? Not at the grammar school we trow. No school but one can instil that, where rules the one best teacher you will ever know, genius though you be— the school kept at your mother's knee. Such mothers as Watt had are the appointed trainers of genius, and make men good and great, if the needed spark be there to enkindle: "Kings they make gods, and meaner subject's kings."

We have another story of Watt's childhood that proclaims the coming man. Precocious children are said rarely to develop far in later years, but Watt was pre-eminently a precocious child, and of this several proofs are related. A friend looking at the child of six said to his father, "You ought to send your boy to a public school, and not allow him to trifle away his time at home." "Look how he is occupied before you condemn him," said the father. He was trying to solve a problem in geometry. His mother had taught him drawing, and with this he was captivated. A few toys were given him, which were constantly in use. Often he took them to pieces, and out of the parts sometimes constructed new ones, a source of great delight. In this way he employed and amused himself in the many long days during which he was confined to the house by ill health.

It is at this stage the steam and kettle story takes its rise. Mrs. Campbell, Watt's cousin and constant companion, recounts, in her memoranda, written in 1798:

Sitting one evening with his aunt, Mrs. Muirhead, at the tea-table, she said: "James Watt, I never saw such an idle boy; take a book or employ yourself usefully; for the last hour you have not spoken one word, but taken off the lid of that kettle and put it on again, holding now a cup and now a silver spoon over the steam, watching how it rises from the spout, and catching and connecting the drops of hot water it falls into. Are you not ashamed of spending your time in this way?"

To what extent the precocious boy ruminated upon the phenomenon must be left to conjecture. Enough that the story has a solid foundation upon which we can build. This more than justifies us in classing it with "Newton and the Apple," "Bruce and the Spider," "Tell and the Apple," "Galvani and the Frog," "Volta and the Damp Cloth," "Washington and His Little Hatchet," a string of gems, amongst the most precious of our legendary possessions. Let no rude iconoclast attempt to undermine one of them. Even if they never occurred, it matters little. They should have occurred, for they are too good to lose. We could part with many of the actual characters of the flesh in history without much loss; banish the imaginary host of the spirit and we were poor indeed. So with these inspiring legends; let us accept them and add others gladly as they arise, inquiring not too curiously into their origin.

While Watt was still in boyhood, his wise father not only taught him writing and arithmetic, but also provided a set of small tools for him in the shop among the workmen—a wise and epoch-making gift, for young Watt soon revealed such wonderful manual dexterity, and could do such astonishing things, that the verdict of one of the workmen, "Jamie has a fortune at his finger-ends," became a common saying among them. The most complicated work seemed to come naturally to him. One model after another was produced to the wonder and delight of his older fellow-workmen. Jamie was the pride of the shop, and no doubt of his fond father, who saw with pardonable pride that his promising son inherited his own traits, and gave bright promise of excelling as a skilled handicraftsman.

The mechanical dexterity of the Watts, grandfather, father and son, is not to be belittled, for most of the mechanical inventions have come from those who have been cunning of hand and have worked as manual laborers, generally in charge of the machinery or devices which they have improved. When new processes have been invented, these also have usually suggested themselves to the able workmen as they experienced the crudeness of existing methods. Indeed, few important inventions have come from those who have not been thus employed. It is with inventors as with poets; few have been born to the purple or with silver spoons in their mouths, and we shall plainly see later on that had it not been for Watt's inherited and acquired manual dexterity, it is probable that the steam engine could never have been perfected, so often did failure of experiments arise solely because it was in that day impossible to find men capable of executing the plans of the inventor. His problem was to teach them by example how to obtain the exact work required when the tools of precision of our day were unknown and the men themselves were only workmen of the crudest kind. Many of the most delicate parts, even of working engines, passed through Watt's own hands, and for most of his experimental devices he had himself to make the models. Never was there an inventor who had such reason to thank fortune that in his youth he had learned to work with his hands. It proved literally true, as his fellow-workmen in the shop predicted, that "Jamie's fortune was at his finger-ends."

As before stated, he proved a backward scholar for a time, at the grammar school. No one seems to have divined the latent powers smoldering within. Latin and Greek classics moved him not, for his mind was stored with more entrancing classics learned at his mother's knee: his heroes were of nobler mould than the Greek demigods, and the story of his own romantic land more fruitful than that of any other of the past. Busy working man has not time to draw his inspiration from more than one national literature. Nor has any man yet drawn fully from any but that of his native tongue. We can no more draw our mental sustenance from two languages than we can think in two. Man can have but one deep source from whence come healing waters, as he can have but one mother tongue. So it was with Watt. He had Scotland and that sufficed. When the boy absorbs, or rather is absorbed by, Wallace, The Bruce, and Sir John Grahame, is fired by the story of the Martyrs, has at heart page after page of the country's ballads, and also, in more recent times, is at home with Burns' and Scott's prose and poetry, he has little room and less desire, and still less need, for inferior heroes. So the dead languages and their semi-supernatural, quarrelsome, self-seeking heroes passed in review without gaining admittance to the soul of Watt. But the spare that fired him came at last—Mathematics. "Happy is the man who has found his work," says Carlyle. Watt found his when yet a boy at school. Thereafter never a doubt existed as to the field of his labors. The choice of an occupation is a serious matter with most young men. There was never room for any question of choice with young Watt. The occupation had chosen him, as is the case with genius. "Talent does what it can, genius what it must." When the goddess lays her hand upon a mortal dedicated to her shrine, concentration is the inevitable result; there is no room for anything which does not contribute to her service, or rather all things are made contributory to it, and nothing that the devotee sees or reads, hears or feels, but some way or other is made to yield sustenance for the one great, overmastering task. "The gods send thread for a web begun," because the web absorbs everything that comes within reach. So it proved with Watt.

At fifteen, he had twice carefully read "The Elements of Philosophy" (Gravesend), and had made numerous chemical experiments, repeating them again and again, until satisfied of their accuracy. A small electrical machine was one of his productions with which he startled his companions. Visits to his uncle Muirhead at Glasgow were frequent, and here he formed acquaintance with several educated young men, who appreciated his abilities and kindly nature; but the visits to the same kind uncle "on the bonnie, bonnie banks o' Loch Lomond," where the summer months were spent, gave the youth his happiest days. Indefatigable in habits of observation and research, and devoted to the lonely hills, he extended his knowledge by long excursions, adding to his botanical and mineral treasures. Freely entering the cottages of the people, he spent hours learning their traditions, superstitions, ballads, and all the Celtic lore. He loved nature in her wildest moods, and was a true child of the mist, brimful of poetry and romance, which he was ever ready to shower upon his friends. An omniverous reader, in after life he vindicated his practice of reading every book he found, alleging that he had "never yet read a book or conversed with a companion without gaining information, instruction or amusement." Scott has left on record that he never had met and conversed with a man who could not tell him something he did not know. Watt seems to have resembled

Sir Walter, "who spoke to every man he met as if he were a brother"—as indeed he was—one of the many fine traits of that noble, wholesome character. These two foremost Scots, each supreme in his sphere, seem to have had many social traits in common, and both that fine faculty of attracting others.

The only "sport" of the youth was angling, "the most fitting practice for quiet men and lovers of peace," the "Brothers of the Angle," according to Izaak Walton, "being mostly men of mild and gentle disposition." From the ruder athletic games of the school he was debarred, not being robust, and this was a constant source of morbid misery to him, entailing as it did separation from the other boys. The prosecution of his favorite geometry now occupied his thoughts and time, and astronomy also became a fascinating study. Long hours were often spent, lying on his back in a grove near his home, studying the stars by night and the clouds by day.

Watt met his first irreparable loss in 1753, when his mother suddenly died. The relations between them had been such as are only possible between mother and son. Often had the mother said to her intimates that she had been enabled to bear the loss of her daughter only by the love and care of her dutiful son. Home was home no longer for Jamie, and we are not surprised to find him leaving it soon after she who had been to him the light and leading of his life had passed out of it.

Watt now reached his seventeenth year. His father's affairs were greatly embarrassed. It was clearly seen that the two brothers, John and James, had to rely for their support upon their own unaided efforts. John, the elder, some time before this had taken to the sea and been shipwrecked, leaving only James at home. Of course, there was no question as to the career he would adopt. His fortune "lay at his fingers' ends," and accordingly he resolved at once to qualify himself for the trade of a mathematical instrument maker, the career which led him directly in the pathway of mathematics and mechanical science, and enabled him to gratify his unquenchable thirst for knowledge thereof.

Naturally Glasgow was decided upon as the proper place in which to begin, and Watt took up his abode there with his maternal relatives, the Muirheads, carrying his tools with him.

No mathematical instrument maker was to be found in Glasgow, but Watt entered the service of a kind of jack-of-all-trades, who called himself an "optician" and sold and mended spectacles, repaired fiddles, tuned spinets, made fishing-rods and tackle, etc. Watt, as a devoted brother of the angle, was an adept at dressing trout and salmon flies, and handy at so many things that he proved most useful to his employer, but there was nothing to be learned by the ambitious youth.

His most intimate schoolfellow was Andrew Anderson, whose elder brother, John Anderson, was the well-known Professor of natural philosophy, the first toopen classes for the instruction of working-men in its principles. He bequeathed his property to found an institution for this purpose, which is now a college of the university. The Professor came to know young Watt through his brother, and Watt became a frequent visitor at his house. He was given unrestricted access to the Professor's valuable library, in which he spent many of his evenings.

One of the chief advantages of the public school is the enduring friendships boys form there, first in importance through their beneficial influence upon character, and, second, as aids to success in after life. The writer has been impressed by this feature, for great is the number of instances he has known where the prized working-boy or man in position has been able, as additional force was required, to say the needed word of recommendation, which gave a start or a lift upward to a dearly-cherished schoolfellow. It seems a grave mistake for parents not to educate their sons in the region of home, or in later years in colleges and universities of their own land, so that early friendships may not be broken, but grow closer with the years. Watt at all events was fortunate in this respect. His schoolmate, Andrew Anderson, brought into his life the noted Professor, with all his knowledge, kindness and influence, and opened to him the kind of library he most needed.

CHAPTER II

GLASGOW TO LONDON—RETURN TO GLASGOW

Through Professor Muirhead, a kinsman of Watt's mother, he was introduced to many others of the faculty of the university, and, as usual, attracted their attention, especially that of Dr. Dick, Professor of natural philosophy, who strongly advised him to proceed to London, where he could receive better instruction than it was possible to obtain in Scotland at that time. The kind Professor, diviner of latent genius, went so far as to give him a personal introduction, which proved efficient. How true it is that the worthy, aspiring youth rarely goes unrecognised or unaided. Men with kind hearts, wise heads, and influence strong to aid, stand ready at every turn to take modest merit by the hand and give it the only aid needed, opportunity to speak, through results, for itself. So London was determined upon. Fortunately, a distant relative of the Watt family, a sea-captain, was about to set forth upon that then long and toilsome journey. They started from Glasgow June 7, 1755, on horseback, the journey taking twelve days.

The writer's parents often referred to the fact that when the leading linen manufacturer of Dunfermline was about to take the journey to London—the only man in the town then who ever did—special prayers were always said in church for his safety.

The Member of Parliament in Watt's day from the extreme north of Scotland would have consumed nearly twice twelve days to reach Westminster. To-day if the capital of the English-speaking race were in America, which Lord Roseberry says he is willing it should be, if thereby the union of our English-speaking race were secured, the members of the Great Council from Britain could reach Washington in seven days, the members from British Columbia and California, upon the Pacific, in five days, both land and sea routes soon to be much quickened.

Those sanguine prophets who predict the reunion of our race on both sides of the Atlantic can at least aver that in view of the union of Scotland and England, the element of time required to traverse distances to and from the capital is no obstacle, since the most distant points of the new empire, Britain in the east and British Columbia and California in the west, would be reached in less than one-third the time required to travel from the north of Scotland to London at the time of the union. Besides, the telegraph to-day binds the parts together, keeping all citizens informed, and stirring their hearts simultaneously thousands of miles apart—Glasgow to London, 1755, twelve days; 1905, eight hours. Thus under the genius Steam, tamed and harnessed by Watt, the world shrinks into a neighborhood, giving some countenance to the dreamers who may perchance be proclaiming a coming reality. We may continue, therefore, to indulge the hope of the coming "parliament of man, the federation of the world," or even the older and wider prophecy of Burns, that, "It's coming yet for a' that, when man to man the world o'er, shall brithers be for a' that."

There comes to mind that jewel we owe to Plato, which surely ranks as one of the most precious of all our treasures: "We should lure ourselves as with enchantments, for the hope is great and the reward is noble." So with this enchanting dream, better than most realities, even if it be all a dream. Let the dreamers therefore dream on. The world, minus enchanting dreams, would be commonplace indeed, and let us remember this dream is only dreamable because Watt's steam engine is a reality.

After his twelve days on horseback, Watt arrived in London, a stranger in a strange land, unknowing and unknown. But the fates had been kind for, burdened with neither wealth nor rank, this poor would-be skilled mechanic was to have a fair chance by beginning at the bottom among his fellows, the sternest yet finest of all schools to call forth and strengthen inherent qualities, and impel a poor young man to put forth his utmost effort when launched upon the sea of life, where he must either sink or swim, no bladders being in reserve for him.

Our young hero rose to the occasion and soon proved that, Cæsar-like, he could "stem the waves with heart of controversy." Thus the rude school of experience calls forth and strengthens the latent qualities of youth, implants others, and forms the indomitable man, fit to endure and overcome. Here, for the first time, alone in swarming London, not one relative, not one friend, not even an acquaintance, except the kind sea-captain, challenged by the cold world around to do or die, fate called to Watt as it calls to every man who has his own way to make:

"This is Collingtogle ford,And thou must keep thee with thy sword."

When the revelation first rushes upon a youth, hitherto directed by his parents, that, boy no more, he must act for himself, presto! Change! He is a man, he has at last found himself. The supreme test, which proves the man, can come in all its winnowing force only to those born to earn their own support by training themselves to be able to render to society services which command return. This training compels the development of powers which otherwise would probably lie dormant. Scotch boy as Watt was to the core, with the lowland broad, soft accent, and ignorant of foreign literature, it is very certain that he then found support in the lessons instilled at his mother's knee. He had been fed on Wallace and Bruce, and when things looked darkest, even in very early years, his national hero, Wallace, came to mind, and his struggles against fearful odds, not for selfish ends, but for his country's independence. Did Wallace give up the fight, or ever think of giving up? Never! It was death or victory. Bruce and the spider! Did Bruce falter? Never! Neither would he. "Scots wa hae," "Let us do or die," implanted before his teens, has pulled many a Scottish boy through the crises of life when all was dark, as it will pull others yet to come. Altho Burns and Scott had yet to appear, to crystallise Scotland's characteristics and plant the talismanic words into the hearts of young Scots, Watt had a copious supply of the national sentiment, to give him the "stout heart for the stye brae," when manhood arrived. His mother had planted deep in him, and nurtured, precious seed from her Celtic garden, which was sure to grow and bear good fruit.

We are often met with the question, "What is the best possible safeguard for a young man, who goes forth from a pure home, to meet the temptations that beset his path?" Various answers are given, but, speaking that as a Scot, reared as Watt was, the writer believes all the suggested safeguards combined scarcely weigh as much as preventives against disgracing himself as the thought that it would not be only himself he would disgrace, but that he would also bring disgrace upon his family, and would cause father, mother, sister and brother to hang their heads among their neighbors in secluded village, on far-away moor or in lonely glen. The Scotch have strong traces of the Chinese and Japanese religious devotion to "the family," and the filial instinct is intensely strong. The fall of one member is the disgrace of all. Even although Watt's mother had passed, there remained the venerated father in Greenock, and the letters regularly written to him, some of which have fortunately been preserved, abundantly prove that, tho far from home, yet in home and family ties and family duties the young man had his strong tower of defence, keeping him from "all sense of sin or shame." Watt never gave his father reason for one anxious thought that he would in any respect discredit the good name of his forbears.

Many London shops were visited, but the rules of the trade, requiring apprentices to serve for seven years, or, being journeymen, to have served that time, proved an insuperable obstacle to Watt's being employed. His plan was to fit himself by a year's steady work for return to Glasgow, there to begin on his own account. He had not seven years to spend learning what he could learn in one. He would be his own master. Wise young man in this he was. There is not much outcome in the youth who does not already see himself captain in his dreams, and steers his barque accordingly, true to the course already laid down, not to be departed from, under any stress of weather. We see the kind of stuff this young Scotch lad was made of in the tenacity with which he held to his plan. At last some specimens of his work having seemed very remarkable to Mr. John Morgan, mathematical instrument maker, Finch Lane, Cornhill, he agreed to give the conquering young man the desired year's instructions for his services and a premium of twenty pounds, whereupon the plucky fellow who had kept to his course and made port, wrote to his father of his success, praising his master "as being of as good character, both for accuracy in his business, and good morals, as any of his way in London." The order in which this aspiring young man of the world records the virtues will not be overlooked. He then adds, "If it had not been for Mr. Short, I could not have got a man in London that would have undertaken to teach me, as I now find there are not above five or six who could have taught me all I wanted."

Mr. Short was the gentleman to whom Professor Dick's letter of introduction was addressed, who, no more than the Professor himself, nor Mr. Morgan, could withstand the extraordinary youth, whom he could not refuse taking into his service—glad to get him no doubt, and delighted that he was privileged to instruct one so likely to redound to his credit in after years. Thus Watt made his start in London, the twenty pounds premium being duly remitted from home.

Up to this time, Watt had been a charge on his father, but it was very small, for he lived in the most frugal style at a cost of only two dollars per week. In one of his letters to his father he regrets being unable to reduce it below that, knowing that his father's affairs were not prosperous. He, however, was able to obtain some remunerative work on his own account, which he did after his day's task was over, and soon made his position secure as a workman. Specialisation he met with for the first time, and he expresses surprise that "very few here know any more than how to make a rule, others a pair of dividers, and suchlike." Here we see that even at that early day division of labor had won its way in London, though yet unknown in the country. The jack-of-all-trades, the handyman, who can do everything, gives place to the specialist who confines himself to one thing in which practice makes him perfect. Watt's mission saved him from this, for to succeed he had to be master, not of one process, but of all. Hence we find him first making brass scales, parallel-rulers and quadrants. By the end of one month in this department he was able to finish a Hadley quadrant. From this he proceeded to azimuth compasses, brass sectors, theodolites, and other delicate instruments. Before his year was finished he wrote his father that he had made "a brass sector with a French joint, which is reckoned as nice a piece of framing-work as is in the trade," and expressed the hope that he would soon now be able to support himself and be no longer a charge upon him.

It is highly probable that this first tool finished by his own hands brought to Watt more unalloyed pleasure than any of his greater triumphs of later years, just as the first week's wages of youth, money earned by service rendered, proclaiming coming manhood, brings with it a thrill and glow of proud satisfaction, compared with which all the millions of later years are as dross.

Writers upon labor, who have never labored, generally make the profound mistake of considering labor as one solid mass, when the truth is that it contains orders and degrees as distinct as those in aristocracy. The workman skilled beyond his fellows, who is called upon by his superintendent to undertake the difficult job in emergencies, ranks high, and probably enjoys an honorable title, a pet name conferred by his shopmates. Men measure each other as correctly in the workshop as in the professions, and each has his deserved rank. When the right man is promoted, they rally round and enable him to perform wonders. Where favoritism or poor judgment is shown, the reverse occurs, and there is apathy and dissatisfaction, leading to poor results and serious trouble. The manual worker is as proud of his work, and rightly so, as men are in other vocations. His life and thought centre in the shop as those of members of Congress or Parliament centre in the House; and triumph for him in the shop, his world, means exactly the same to him, and appears not less important to his family and friends than what leadership is to the public man, or in any of the professions. He has all their pride of profession, and less vanity than most.

How far this "pride of profession" extends is well illustrated by the Pittsburgh story of the street scrapers at their noon repast. MacCarthy, recently deceased, was the subject of eulogy, one going so far as to assert that he was "the best man that ever scraped a hoe on Liberty Street." To this, one who had aspirations "allowed Mac was a good enough man on plain work, but around the gas-posts he wasn't worth a cent."

A public character, stopping over night with a friend in the country, the maid-of-all-work tells her mistress, after the guest departs, "I have read so much about him, never expecting to see him; little did I think I should have the honor of brushing his boots this morning." Happy girl in her work, knowing that all service is honorable. Even shoe-blacking, we see, has its rewards.

A Highland laird and lady, visiting some of their crofters on the moors, are met and escorted by a delighted wife to her cot. The children and the husband are duly presented. At an opportune moment the proud wife cannot refrain from informing her visitors that "it was Donald himsel' the laird had to send for to thatch the pretty golf-house at the Castle. Donald did all that himsel'," with an admiring glance cast at the embarrassed great man. Donald "sent for by the laird at the Castle" ranks in Donald's circle and in Donald's own heart with the honor of being sent for by His Majesty to govern the empire in Mr. Balfour's circle and in Mr. Balfour's own heart. Ten to one the proud Highland crofter and his circle reap more genuine, unalloyed satisfaction from the message than the lowland statesman and his circle could reap from his. But it made Balfour famous, you say. So was Donald made famous, his circle not quite so wide as that of his colleague—that is all. Donald is as much "uplifted" as the Prime Minister; probably more so. Thus is human nature ever the same down to the roots. Many distinctions, few differences in life. We are all kin, members of the one family, playing with different toys.

So deep down into the ranks of labor goes the salt of pride of profession, preventing rot and keeping all fresh in the main, because on the humblest of the workers there shines the bright ray of hope of recognition and advancement, progress and success. As long as this vista is seen stretching before all is well with labor. There will be friction, of course, between capital and labor, but it will be healthy friction, needed by, and good for, both. There is the higgling of the market in all business. As long as this valuable quality of honest pride in one's work exists, and finds deserved recognition, society has nothing to fear from the ranks of labor. Those who have had most experience with it, and know its qualities and its failing's best, have no fear; on the contrary, they know that at heart labor is sound, and only needs considerate treatment. The kindly personal attention of the employer will be found far more appreciated than even a rise in wages.

Enforced confinement and unremitting labor soon told upon Watt's delicate constitution, yet he persevered with the self-imposed extra work, which brought in a little honest money and reduced the remittances from home. He caught a severe cold during the winter and was afflicted by a racking cough and severe rheumatic pains. With his father's sanction, he decided to return home to recuperate, taking good care however, forehanded as he always proved himself, to secure some new and valuable tools and a stock of materials to make many others, which "he knew he must make himself." A few valuable books were not forgotten, among them Bion's work on the "Construction and Use of Mathematical Instruments"—nothing pertaining to his craft but he would know. King he would be in that, so everything was made to revolve around it. That was the foundation upon which he had to build.

To the old home in Scotland our hero's face was now turned in the autumn of 1756, his twentieth year. His native air, best medicine of all for the invalid exile, soon restored his health, and to Glasgow he then went, in pursuance of his plan of life early laid down, to begin business on his own account. He thus became master before he was man. There was not in all Scotland a mathematical instrument maker, and here was one of the very best begging permission to establish himself in Glasgow. As in London so in Glasgow, however, the rules of the Guild of Hammermen, to which it was decided a mathematical instrument maker would belong, if one of such high calling made his appearance, prevented Watt from entrance if he had not consumed seven years in learning the trade. He had mastered it in one, and was ready to demonstrate his ability to excel by any kind of test proposed. Watt had entered in properly by the door of knowledge and experience of the craft, the only door through which entrance was possible, but he had travelled too quickly; besides he was "neither the son of a burgess, nor had he served an apprenticeship in the borough," and this was conclusive. How the world has travelled onward since those days! and yet our day is likely to be in as great contrast a hundred and fifty years hence. Protective tariffs between nations, and probably wars, may then seem as strangely absurd as the hammermen's rules. Even in 1905 we have still a far road to travel.

Failing in his efforts to establish himself in business, he asked the guild to permit him to rent and use a small workshop to make experiments, but even this was refused. We are disposed to wonder at this, but it was in strict accordance with the spirit of the times.

When the sky was darkest, the clouds broke and revealed the university as his guardian angel. Dr. Dick, Professor of natural philosophy, knowing of Watt's skill from his first start in Glasgow, had already employed him to repair some mathematical instruments bequeathed to the university by a Scotch gentleman in the West Indies, and the work had been well done, at a cost of five pounds—the first contract money ever earned by Watt in Glasgow. Good work always tells. Ability cannot be kept down forever; if crushed to earth, it rises again. So Watt's "good work" brought the Professors to his aid, several of whom he had met and impressed most favorably during its progress. The university charter, gift of the Pope in 1451, gave absolute authority within the area of its buildings, and the Professors resolved to give our hero shelter there—the best day's work they ever did. May they ever be remembered for this with feelings of deepest gratitude? What men these were! The venerable Anderson has already been spoken of; Adam Smith, who did for the science of economics what Watt did for steam, was one of Watt's dearest friends; Black, discoverer of latent heat; Robinson, Dick of whom we have spoken, and others. Such were the world's benefactors, who resolved to take Watt under their protection, and thus enabled him to do his appointed work. Glorious university, this of Glasgow, protector and nurse of Watt, probably of all its decisions this has been of the greatest service to man!

There are universities and universities. Glasgow's peculiar claim to regard lies in the perfect equality of the various schools, the humanities not neglected, the sciences appreciated, neither accorded precedence. Its scientific Professor, Thompson, now Lord Kelvin, was recently elevated to the Lord Chancellorship, the highest honor in its power to bestow.

Every important university develops special qualities of its own, for which it is noted. That of Glasgow is renowned for devotion to the scientific field. What a record is hers! Protector of Watt, going to extreme measures necessary, not alone to shelter him, but to enable him to labor within its walls and support himself; first university to establish an engineering school and professorship of engineering; first to establish a chemical teaching laboratory for students; first to have a physical laboratory for the exercise and instruction of students in experimental work; nursery from which came the steam engine of Watt, the discovery of latent heat by its Professor Black, and the successful operation of telegraph cables by its Professor and present Lord Chancellor (Lord Kelvin). May the future of Glasgow University copy fair her glorious past! Her "atmosphere" favors and stimulates steady, fruitful work. At all Scottish, as at all American universities, we may rejoice that there is always found a large number of the most distinguished students, who, figuratively speaking, cultivate knowledge upon a little oatmeal, earning money between terms to pay their way. It is highly probable that a greater proportion of these will be heard from in later years than of any other class.

American universities have, fortunately, followed the Glasgow model, and are giving more attention to the hitherto much neglected needs of science, and the practical departments of education, making themselves real universities, "where any man can study everything worth studying."

A room was assigned to Watt, only about twenty feet square, but it served him as it has done others since for great work. When the well-known author, Dr. Smiles, visited the room, he found in it the galvanic apparatus employed by Professor Thompson (Lord Kelvin) for perfecting his delicate invention which rendered ocean cables effective.

The kind and wise Professors did not stop here. They went pretty far, one cannot but think, when they took the next step in Watt's behalf, giving him a small room, which could be made accessible to the public, and this he was at liberty to open as a shop for the sale of his instruments, for Watt had to make a living by his handiwork. Strange work this for a university, especially in those days; but our readers, we are sure, will heartily approve the last, as they have no doubt approved the first action of the faculty in favor of struggling genius. Business was not prosperous at first with Watt, his instruments proving slow of sale. Of quadrants he could make three per week with the help of a lad, at a profit of forty shillings, but as sea-going ships could not then reach Glasgow, few could be sold. A supply was sent to Greenock, then the port of Glasgow, and sold by his father. He was reduced, as the greatest artists have often been, to the necessity of making what are known as "pot-boilers." Following the example of his first master in Glasgow he made spectacles, fiddles, flutes, guitars, and, of course, flies and fishing-tackle, and, as the record tells, "many dislocated violins, fractured guitars, fiddles also, if intreated, did he mend with good approbation." Such were his "pot-boilers" that met the situation.

His friend, Professor Black, who, like Professor Dick, had known of Watt's talent, one day asked him if he couldn't make an organ for him. By this time, Watt's reputation had begun to spread, and it finally carried him to the height of passing among his associates as "one who knew most things and could make anything." Watt knew nothing about organs, but he immediately undertook the work (1762), and the result was an indisputable success that led to his constructing, for a mason's lodge in Glasgow, a larger "finger organ," "which elicited the surprise and admiration of musicians." This extraordinary man improved everything he touched. For his second organ he devised a number of novelties, a sustained monochord, indicators and regulators of the blast, means for tuning to any system, contrivances for improving the stops, etc.

Lest we are led into a sad mistake here, let us stop a moment to consider how Watt so easily accomplished wonders, as if by inspiration. In all history it may be doubted whether success can be traced more clearly to long and careful preparation than in Watt's case. When we investigate, for instance, this seeming sleight-of-hand triumph with the organs, we find that upon agreeing to make the first, Watt immediately devoted himself to a study of the laws of harmony, making science supplement his lack of the musical ear. As usual, the study was exhaustive. Of course he found and took for guide the highest authority, a profound, but obscure book by Professor Smith of Cambridge University, and, mark this, he first made a model of the forthcoming organ. It is safe to say that there was not then a man in Britain who knew more of the science of music and was more thoroughly prepared to excel in the art of making organs than the new organ-builder.

When he attacked the problem of steam, as we shall soon see, the same course was followed, although it involved the mastering of three languages, that he should miss nothing.

We note that the taking of infinite pains, this fore-arming of himself, this knowing of everything that was to be known, the note of thorough preparation in Watt's career, is ever conspicuous. The best proof that he was a man of true genius is that he first made himself master of all knowledge bearing upon his tasks.

Watt could not have been more happily situated. His surroundings were ideal, the resources of the university were at his disposal, and, being conveniently situated, his workshop soon became the rendezvous of the faculty. He thus enjoyed the constant intimate companionship of one of the most distinguished bodies of educated men of science in the world. Glasgow was favored in her faculty those days as now. Two at least of Watt's closest friends, the discoverer of latent heat, and the author of the "Wealth of Nations," won enduring fame. Others were eminent. He did not fail to realise his advantages, and has left several acknowledgments of his debt to "those who were all much my superiors, I never having attended a college and being then but a mechanic." His so-called superiors did not quite see it in this light, as they have abundantly testified, but the modesty of Watt was ever conspicuous all through his life.

Watt led a busy life, the time not spent upon the indispensable "pot-boilers" being fully occupied in severe studies; chemistry, mathematics and mechanics all received attention. What he was finally to become no one could so far predict, but his associates expected something great from one who had so deeply impressed them.

Robison (afterwards Professor of natural history in Edinburgh University), being nearer Watt's age than the others, became his most intimate friend. His introduction to Watt, in 1758, has been described by himself. After feasting his eyes on the beautifully finished instruments in his shop, Robison entered into conversation with him. Expecting to find only a workman, he was surprised to find a philosopher. Says Robison:

I had the vanity to think myself a pretty good proficient in my favorite study (mathematical and mechanical philosophy), and was rather mortified at finding Mr. Watt so much my superior. But his own high relish for those things made him pleased with the chat of any person who had the same tastes with himself; or his innate complaisance made him indulge my curiosity, and even encourage my endeavors to form a more intimate acquaintance with him. I lounged much about him, and, I doubt not, was frequently teasing him. Thus our acquaintance began.

CHAPTER III

Captured by Steam

The supreme hour of Watt's life was now about to strike. He had become deeply interested in the subject of steam, to which Professor Robison had called his attention, Robison being then in his twentieth year, Watt three years older.

Robison's idea was that steam might be applied to wheel carriages. Watt admitted his ignorance of steam then. Nevertheless, he made a model of a wheel carriage with two cylinders of tin plate, but being slightly and inaccurately made, it failed to work satisfactorily. Nothing more was heard of it. Robison soon thereafter left Glasgow. The demon Steam continued to haunt Watt. He, who up to this time had never seen even a model of a steam engine, strangely discovered in his researches that the university actually owned a model of the latest type, the Newcomen engine, which had been purchased for the use of the natural philosophy class. One wonders how many of the universities in Britain had been so progressive. That of Glasgow seems to have recognised at an early day the importance of science, in which department she continues famous. The coveted and now historical model had been sent to London for repairs. Watt urged its prompt return and a sum of money was voted for this purpose. Watt was at last completely absorbed in the subject of steam. He read all that had been written on the subject. Most of the valuable matter those days was in French and Italian, of which there were no translations. Watt promptly began to acquire these languages that he might know all that was to be known. He could not await the coming of the model, which did not arrive until 1763, and began his own experiments in 1761. How he obtained the necessary appliances and apparatus, one asks. The answer is easy. He made them. Apothecaries' vials were his steam boilers, and hollowed-out canes his steam-pipes. Numerous experiments followed and much was learnt. Watt's account of these is appended to the article on "Steam and the Steam Engine" in the "Encyclopædia Britannica," ninth edition.

Detailed accounts of Watt's numerous experiments, failures, difficulties, disappointments, and successes, as one after the other obstacles were surmounted, is not within the scope of this volume, these being all easily accessible to the student, but the general reader may be interested in the most important of all the triumphs of the indefatigable worker—the keystone of the arch. The Newcomen model arrived at last and was promptly repaired, but was not successful when put in operation. Steam enough could not be obtained, although the boiler seemed of ample capacity. The fire was urged by blowing and more steam generated, and still it would not work; a few strokes of the piston and the engine stopped. Smiles says that exactly at the point when ordinary experimentalists would have abandoned the task, Watt became thoroughly aroused. "Every obstacle," says Professor Robison, "was to him the beginning of a new and serious study, and I knew he would not quit it until he had either discovered its worthlessness or had made something of it." The difficulty here was serious. Books were searched in vain. No one had touched it. A course of independent experiments was essential, and upon this he entered as usual, determined to find truth at the bottom of the well and to get there in his own way. Here he came upon the fact which led him to the stupendous result. That fact was the existence of latent heat, the original discoverer of which was Watt's intimate friend, Professor Black. Watt found that water converted into steam heated five times its own weight of water to steam heat. He says:

Being struck with this remarkable fact (effect of latent heat), and not understanding the reason of it, I mentioned it to my friend, Dr. Black, who then explained to me his doctrine of latent heat, which he had taught some time before this period (1764); but having myself been occupied with the pursuits of business, if I had heard of it I had not attended to it, when I thus stumbled upon one of the material facts by which that beautiful theory is supported.

Here we have an instance of two men in the same university, discovering latent heat, one wholly ignorant of the other's doings; fortunately, the later discoverer only too glad to acknowledge and applaud the original, and, strange to say, going to him to announce the discovery he had made. Watt of course had no access to the Professor's classes, and some years before the former stumbled upon the fact, the theory had been announced by Black, but had apparently attracted little attention. This episode reminds us of the advantages Watt had in his surroundings. He breathed the very "atmosphere" of scientific and mechanical investigation and invention, and had at hand not only the standard books, but the living men who could best assist him.

What does latent heat mean? We hear the reader inquire. Let us try to explain it in simple language. Arago pronounced Black's experiment revealing it as one of the most remarkable in modern physics. Water passed as an element until Watt found it was a compound. Change its temperature and it exists in three different states, liquid, solid, and gaseous—water, ice and steam. Convert water into steam, and pass, say, two pounds of steam into ten pounds of water at freezing point and the steam would be wholly liquified, *i.e.*, become water again, at 212°, but the whole ten pounds of freezing water would also be raised to 212° in the process. That is to say two pounds of steam will convert ten pounds of freezing water into boiling water, so great is the latent heat set free in the passage of steam to lower temperatures at the moment when the contact of cold surfaces converts the vapor from the gaseous into the liquid state. This heat is so thoroughly merged in the compound that the most delicate thermometer cannot detect a variation. It is undiscoverable by our senses and yet it proves its existence beyond question by its work. Heat which is obtained by the combustion of coal or wood, lies also in water, to be drawn forth and utilised in steam. It is apparently a mere question of temperature. The heat lies latent and dead until we raise the temperature of the water to 212°, and it is turned to vapor. Then the powerful force is instantly imbued with life and we harness it for our purposes.

The description of latent heat which gave the writer the clearest idea of it, and at the same time a much-needed reminder of the fact that Watt was the discoverer of the practically constant and unvarying amount of heat in steam, whatever the pressure, is the following by Mr. Lauder, a graduate of Glasgow University and pupil of Lord Kelvin, taken from "Watt's Discoveries of the Properties of Steam."

It is well to distinguish between the two things, Discovery and Invention. The title of Watt the Inventor is world-wide, and is so just and striking that there is none to gainsay. But it is only to the few that dive deeper that Watt the Discoverer is known. When his mind became directed to the possibilities of the power of steam, he, following his natural bent, began to investigate its properties. The mere inventor would have been content with what was already known, and utilised such knowledge, as Newcomen had done in his engine. Watt might have invented the separate condenser and ranked as a great inventor, but the spirit of enquiry was in possession of him, and he had to find out all he could about the *nature* of steam.

His first discovery was that of latent heat. When communicating this to Professor Black he found that his friend had anticipated him, and had been teaching it in lectures to his students for some years past. His next step was the discovery of the *total* heat of steam, and that this remains practically constant at all pressures. Black's fame rests upon his theory of latent heat;

Watt's fame as the discoverer of the total heat of steam should be equally great, and would be no doubt had his rôle of inventor not overshadowed all his work.

This part of Watt's work has been so little known that it is almost imperative to-day to give some idea of it to the general reader. Suppose you take a flask, such as olive oil is often sold in, and fill with cold water. Set it over a lighted lamp, put a thermometer in the water, and the temperature will be observed to rise steadily till it reaches 212°, where it remains, the water boils, and steam is produced freely. Now draw the thermometer out of the water, but leaving it still in the steam. It remains steady at the same point—212°. Now it requires quite a long time and a large amount of heat to convert all the water into steam. As the steam goes off at the same temperature as the water, it is evident a quantity of heat has escaped in the steam, of which the thermometer gives us no account. This is latent heat.

Now, if you blow the steam into cold water instead of allowing it to pass into the air, you will find that it heats the water six times more than what is due to its indicated temperature. To fix your ideas: suppose you take 100 lbs. of water at 60°, and blow one pound of steam into it, making 101 lbs., its temperature will now be about 72°, a rise of 12°. Return to your 100 lbs. of water at 60° and add one pound of water at 212° the same temperature as the steam you added, and the temperature will only be raised about 2°. The one pound of steam heats six times more than the one pound of water, both being at the same temperature. This is the quantity of latent heat, which means simply hidden heat, in steam.

Proceeding further with the experiment, if, instead of allowing the steam to blow into the water, you confine it until it gets to some pressure, then blow it into the water, it takes the same weight to raise the temperature to the same degree. This means that the total heat remains practically the same, no matter at what pressure.

This is James Watt's discovery, and it led him to the use of high-pressure steam, used expansively.

Even coal may yet be superseded before it is exhausted, for as eminent an authority as Professor Pritchett of the Massachusetts Institute of Technology has said in a recent address:

Watt's invention and all it has led to is only a step on the way to harnessing the forces of nature to the service of man. Do you doubt that other inventions will work changes even more sweeping than those which the steam engine has brought?

Consider a moment. The problem of which Watt solved a part is not the problem of inventing a machine, but the problem of using and storing the forces of nature which now go to waste. Now to us who live on the earth there is only one source of power—the sun. Darken the sun and every engine on the earth's surface would soon stop, every wheel cease to turn, and all movement cease. How prodigal this supply of power is we seldom stop to consider. Deducting the atmospheric absorption, it is still true that the sun delivers on each square yard of the earth's surface, when he is shining, the equivalent of one horse-power working continuously. Enough mechanical power goes to waste on the college campus to warm and light and supply all the manufactories, street railroads and other consumers of mechanical power in the city. How to harness this power and to store it—that is the problem of the inventor and the engineer of the twentieth century, a problem which in good time is sure to be solved.

Who shall doubt, after finding this secret source of force in water, that some future Watt is to discover other sources of power, or perchance succeed in utilising the superabundant power known to exist in the heat of the sun, or discover the secret of the latent force employed by nature in animals, which converts chemical energy directly into the dynamic form, giving much higher efficiencies than any thermo-dynamic machine has to-day or probably ever can have. Little knew Shakespeare of man's perfect power of motion which utilises all energy! How came he then to exclaim "What a piece of work is man; how infinite in faculty; in form and *moving* how express and admirable"? This query, and a thousand others, have arisen; for we forget Arnold's lines to the Master:

"Others abide our question. Thou art free.We ask and ask—thou smilest and art still."

Man's "moving" is found more "express and admirable" than that of the most perfect machine or adaptation of natural forces yet devised. Lord Kelvin says the animal motor more closely resembles an electro-magnetic engine than a heat engine, but very probably the chemical forces in animals produce the external mechanical effects through electricity and do not act as a thermo-dynamic engine.

The wastage of heat energy under present methods is appalling. About 65 per cent. of the heat energy of coal can be put into the steam boiler, and from this only 15 per cent. of mechanical power is obtained. Thus about nine-tenths of the original heat in coal is wasted. Proceeding further and putting mechanical power into electricity, only from 2 to 5 per cent. is turned into light; or, in other words, from coal to light we get on an average only about one-half of 1 per cent. of the original energy, a wastage of ninety-nine and one-half of every hundred pounds of coal used. The very best possible with largest and best machinery is a little more than one pound from every hundred consumed.

When Watt gave to the steam-engine five times its efficiency by utilising the latent heat, he only touched the fringe of the mysterious realm which envelops man.

Burbank, of the spineless cactus and new fruits, who has been delving deep into the mysteries, tells us:

The facts of plant life demand a kinetic theory of evolution, a slight change from Huxley's statement that, "Matter is a magazine of force," to that of matter being force alone. The time will come when the theory of "ions" will be thrown aside, and no line left between force and matter.

Professor Matthews, he who, with Professor Loeb at Wood's Hole, is imparting life to sea-urchins through electrical reactions, declares "that certain chemical substances coming together under certain conditions are bound to produce life. All life comes through the operation of universal laws." We are but young in all this mysterious business. What lies behind and probably near at hand may not merely revolutionise material agencies but human preconceptions as well. "There are more things in Heaven and Earth than are ever dreamt of in your Philosophy."

Latent Heat was a find indeed, but there remained another discovery yet to make. Watt found that no less than four-fifths of all the steam used was lost in heating the cold cylinder, and only one-fifth performed service by acting on the piston. Prevent this, and the power of the giant is increased fourfold. Here was the prize to contend for. Win this and the campaign is won. First then, what caused the loss? This was soon determined. The cylinder was necessarily cooled at the top because it was open to the air, and also cooled below in condensing the charge of steam that had driven the piston up in order to create a vacuum, without which the piston would not descend from top to bottom, to begin another upward stroke. A jet of cold water was introduced to effect this. How to surmount this seemingly insuperable obstacle was the problem that kept Watt long in profound study.

Many plans were entertained, only to be finally rejected. At last the flash came into that teeming brain like a stroke of lightning. Eureka! he had found it. Not one scintilla of doubt ever intruded thereafter. The solution lay right there and he would invent the needed appliances. His mode of procedure, when on the trail of big game, is beautifully illustrated here. When he found the root of the defect which rendered the Newcomen engine impracticable for general purposes, he promptly formulated the one indispensable condition which alone met the problem, and which the successful steam-engine must possess. He abandoned all else for the time as superfluous, since this was the key of the position. This is the law he then laid down as an axiom—which is repeated in his specification for his first patent in 1769: "To make a perfect steam engine it was necessary that the cylinder should be always as hot as the steam which entered it, and that the steam should be cooled below 100° to exert its full powers."

Watt describes how at last the idea of the "separate condenser," the complete cure, flashed suddenly upon his mind:

I had gone to take a walk on a fine Sabbath afternoon, early in 1765. I had entered the green by the gate at the foot of Charlotte Street and had passed the old washing-house. I was thinking upon the engine at the time, and had gone as far as the herd's house, when the idea came into my mind that as steam was an elastic body it would rush into a vacuum, and if a communication were made between the cylinder and an exhausted vessel it would rush into it, and might be there condensed without cooling the cylinder. I then saw that I must get rid of the condensed steam and injection-water if I used a jet as in Newcomen's engine. Two ways of doing this occurred to me. First, the water might be run off by a descending pipe, if an offlet could be got at the depth of thirty-five or thirty-six feet, and any air might be extracted by a small pump. The second was to make the pump large enough to extract both water and air ... I had not walked farther than the golf-house when the whole thing was arranged in my mind.

Professor Black says, "This capital improvement flashed upon his mind at once and filled him with rapture." We may imagine

"Then felt he like some watcher of the skiesWhen a new planet sweeps into his ken."

A new world had sprung forth in Watt's brain, for nothing less has the steam engine given to man. One reads with a smile the dear modest man's deprecatory remarks about the condenser in after years, when he was overcome by the glowing tributes paid him upon one occasion and hailed as having conquered hitherto uncontrollable steam. He stammered out words to the effect that it came in his way and he happened to find it; others had missed it; that was all; somebody had to stumble upon it. That is all very well, and we love thee, Jamie Watt (he was always Jamie to his friends), for such self-abnegation, but the truth of history must be vindicated for all that. It proclaims, Thou art the man; go up higher and take your seat there among the immortals, the inventor of the greatest of all inventions, a great discoverer and one of the noblest of men!

In this one change lay all the difference between the Newcomen engine, limited to atmospheric pressure, and the steam engine, capable of development into the modern engine through the increasing use of the tremendous force of steam under higher pressures, and improved conditions from time to time.

Watt leads the steam out of the cylinder and condenses it in a separate vessel, leaving the cylinder hot. He closes the cylinder top and sends a circular piston (hitherto all had been square) through it, and closely stuffs it around to prevent escape of steam. The rapidity of the "strokes" gained keeps the temperature of the cylinder high; besides, he encases it and leaves a space between cylinder and covering filled with steam. Thus he fulfils his law: "The cylinder is kept as hot as the steam that enters." "How simple!" you exclaim. "Is that all? How obviously this is the way to do it!" Very true, surprised reader, but true, also, that no condenser and closed cylinder, no modern steam engine.

On Monday morning following the Sabbath flash, we find Watt was up betimes at work upon the new idea. How many hours' sleep he had enjoyed is not recorded, but it may be imagined that he had several visions of the condenser during the night. One was to be made at once; he borrowed from a college friend a brass syringe, the body of which served as a cylinder. The first condenser vessel was an improvised syringe and a tin can. From such an acorn the mighty oak was to grow. The experiment was successful and the invention complete, but Watt saw clearly that years of unceasing labor might yet pass before the details could all be worked out and the steam engine appear ready to revolutionise the labor of the world. During these years, Professor Black was his chief adviser and encouraged him in hours of disappointment. The true and able friend not only did this, but furnished him with money needed to enable him to concentrate all his time and strength upon the task.

Most opportunely, at this juncture, came Watt's marriage, to his cousin Miss Miller, a lady to whom he had long been deeply attached. Watt's friends are agreed in stating that the marriage was of vast importance, for he had not passed untouched through the days of toil and trial. Always of a meditative turn, somewhat prone to melancholy when without companionship, and withal a sufferer from nervous headaches, there was probably no gift of the gods equal to that of such a wife as he had been so fortunate as to secure. Gentle yet strong in her gentleness, it was her courage, her faith, and her smile that kept Watt steadfast. No doubt he, like many other men blessed with an angel in the household, could truly aver that his worrying cares vanished at the doorstep.

Watt had at last, what he never had before, a home. More than one intimate friend has given expression to the doubt whether he could have triumphed without Mrs. Watt's bright and cheerful temperament to keep him from despondency during the trying years which he had now to encounter. Says Miss Campbell:

I have not entered into any of the interesting details my mother gave me of Mr. Watt's early and constant attachment to his cousin Miss Miller; but she ever considered it as having added to his enjoyment of life, and as having had the most beneficial influence on his character. Even his powerful mind sank occasionally into misanthropic gloom, from the pressure of long-continued nervous headaches, and repeated disappointments in his hopes of success in life. Mrs. Watt, from her sweetness of temper, and lively, cheerful disposition, had power to win him from every wayward fancy; to rouse and animate him to active exertion. She drew out all his gentle virtues, his native benevolence and warm affections.

From all that has been recorded of her, we are justified in classing Watt with Bassanio.

"It is very meetHe live an upright life,For having such a blessing in his lady,He finds the joys of heaven here on earth;And if on earth he do not merit it,In reason he should never come to heaven."

Watt knew and felt this and let us hope that, as was his duty, he let Mrs. Watt know it, not only by act, but by frequent acknowledgment.

Watt did not marry imprudently, for his instrument-making business had increased, as was to have been expected, for his work soon made a reputation as being most perfectly executed. At first he was able to carry out all his orders himself; now he had as many as sixteen workmen. He took a Mr. Craig as a partner, to obtain needed capital. His profits one year were $3,000. The business had been removed in 1760 to new quarters in the city, and Watt himself had rented a house outside the university grounds. Having furnished it, Watt brought his young wife and installed her there, July, 1764. We leave him there, happy in the knowledge that he is to be carefully looked after, and, last but not least, steadily encouraged and counselled not to give up the engine. As we shall presently see, such encouragement was much needed at intervals.

The first step was to construct a model embodying all the inventions in a working form. An old cellar was rented, and there the work began. To prepare the plan was easy, but its execution was quite another story. Watt's sad experience with indifferent work had not been lost upon him, and he was determined that, come what may, this working model should not fail from imperfect construction. His own handiwork had been of the finest and most delicate kind, but, as he said, he had "very little experience of mechanics *in great.*" This model was a monster in those days, and great was the difficulty of finding mechanics capable of carrying out his designs. The only available men were blacksmiths and tinsmiths, and these were most clumsy workmen, even in their own crafts. Were Watt to revisit the earth to-day, he would not easily find a more decided change or advance over 1764, in all that has been changed or improved since then, than in this very department of applied mechanics. To-day such a model as Watt constructed in the cellar would be simple work indeed. Even the gasoline or the electric motor of to-day, though complicated far beyond the steam model, is now produced by automatic machinery. Skilled workmen do not have to fashion the parts. They only stand looking on at machinery—itself made by automatic tools—performing work of unerring accuracy. Had Watt had at his call only a small part of the inventory resources of our day, his model steam engine might have been named the Minerva, for Minerva-like, it would have sprung forth complete, the creature of automatic machinery, the workmen meanwhile smilingly looking on at these slaves of the mechanic which had been brought forth and harnessed to do his bidding by the exercise of godlike reason.

The model was ready after six months of unceasing labor, but notwithstanding the scrupulous fastidiousness displayed by Watt in the workmanship of all the parts, the machine, alas, "snifted at many openings." Little can our mechanics of to-day estimate what "perfect joints" meant in those days. The entire correctness of the great idea was, however, demonstrated by the trials made.

The right principle had been discovered; no doubt of that. Watt's decision was that "it must be followed to an issue." There was no peace for him otherwise. He wrote (April, 1765) to a friend, "My whole thoughts are bent on this machine. I can think of nothing else." Of course not; he was hot in the chase of the biggest game hunter ever had laid eyes on. He had seen it, and he knew he had the weapons to bring it down. A larger model, free as possible from defects which he felt he could avoid in the next, was promptly determined upon. A larger and better shop was obtained, and here Watt shut himself up with an assistant and erected the second model. Two months sufficed, instead of six required for the first. This one also at first trial leaked in many directions, and the condenser needed alterations. Nevertheless, the engine accomplished much, for it worked readily with ten and one-half pounds pressure per square inch, a decided increase over previous results. It was still the cylinder and its piston that gave Watt the chief trouble. No wonder the cylinder leaked. It had to be hammered into something like true lines, for at that day so backward was the art that not even the whole collective mechanical skill of cylinder-making could furnish a bored cylinder of the simplest kind. This is not to be construed as unduly hard upon Glasgow, for it is said that all the skill of the world could not do so in 1765, only one hundred and forty years ago. We travel so fast that it is not surprising that there are wiseacres among us quite convinced that we are standing still.

We may be pardoned for again emphasising the fact that it is not only for his discoveries and inventions that Watt is to be credited, but also for the manual ability displayed in giving to these "airy nothings of the brain, a local habitation and a name," for his greatest idea might have remained an "airy nothing," had he not been also the mechanician able to produce it in the concrete. It is not, therefore, only Watt the inventor, Watt the discoverer, but also Watt, the manual worker, that stands forth. As we shall see later on, he created a new type of workmen capable of executing his plans, working with, and educating them often with his own hands. Only thus did he triumph, laboring mentally and physically. Watt therefore must always stand among the benefactors of men, in the triple capacity of discoverer, inventor, and constructor.

The defects of the cylinder, though serious, were clearly mechanical. Their certain cure lay in devising mechanical tools and appliances and educating workmen to meet the new demands. An exact cylinder would leave no room for leakage between its smooth and true surface and the piston; but the solution of another difficulty was not so easily indicated. Watt having closed the top of the cylinder to save steam, was debarred from using water on the upper surface of the piston as Newcomen did, to fill the interstices between piston and cylinder and prevent leakage of steam, as his piston was round and passed through the top of the cylinder. The model leaked badly from this cause, and while engaged trying numerous expedients to meet this, and many different things for stuffing, he wrote to a friend, "My old White Iron man is dead." This being the one he had trained to be his best mechanic, was a grievous loss in those days. Misfortunes never come singly; he had just started the engine after overhauling it, when the beam broke. Discouraged, but not defeated, he battled on, steadily gaining ground, meeting and solving one difficulty after another, certain that he had discovered how to utilise steam.

CHAPTER IV

PARTNERSHIP WITH ROEBUCK

Capital was essential to perfect and place the engine upon the market; it would require several thousand pounds. Had Watt been a rich man, the path would have been clear and easy, but he was poor, having no means but those derived from his instrument-making business, which for some time had necessarily been neglected. Where was the daring optimist who could be induced to risk so much in an enterprise of this character, where result was problematical. Here, Watt's best friend, Professor Black, who had himself from his own resources from time to time relieved Watt's pressing necessities, proved once more the friend in time of need. Black thought of Dr. Roebuck, founder of the celebrated Carron Iron Works near by, which Burns apostrophised in these lines, when denied admittance:

"We cam Na here to view your worksIn hopes to be mair wise, But only lest we gang to hellIt may be nae surprise."

He was approached upon the subject by Dr. Black, and finally, in September, 1765, he invited Watt to visit him with the Professor at his country home, and urged him to press forward his invention "whether he pursued it as a philosopher or as a man of business." In the month of November Watt sent Roebuck drawings of a covered cylinder and piston to be cast at his works, but it was so poorly done as to be useless. "My principal difficulty in making engines," he wrote Roebuck, "is always the smith-work."

By this time, Watt was seriously embarrassed for money. Experiments cost much and brought in nothing. His duty to his family required that he should abandon these for a time and labor for means to support it. He determined to begin as a surveyor, as he had mastered the art when making surveying instruments, as was his custom to study and master wherever he touched. He could never rest until he knew all there was to know about anything. Of course he succeeded. Everybody knew he would, and therefore business came to him. Even a public body, the magistrates of Glasgow, had not the slightest hesitation in obtaining his services to survey a canal which was to open a new coal field. He was also commissioned to survey the proposed Forth and Clyde canal. Had he been content to earn money and become leading surveyor or engineer of Britain, the world might have waited long for the forthcoming giant destined to do the world's work; but there was little danger of this. The world had not a temptation that could draw Watt from his appointed work. His thoughts were ever with his engine, every spare moment being devoted to it. Roebuck's speculative and enterprising nature led him also into the entrancing field of steam. It haunted him until finally, in 1767, he decided to pay off Watt's debts to the amount of a thousand pounds, provide means for further experiments, and secure a patent for the engine. In return, he became owner of two thirds of the invention.

Next year Watt made trial of a new and larger model, with unsatisfactory results upon the first trial. He wrote Roebuck that "by an unforeseen misfortune, the mercury found its way into the cylinder and played the devil with the solder." Only after a month's hard labor was the second trial made, with very different and indeed astonishing results—"success to my heart's content," exclaimed Watt. Now he would pay his long-promised debt to his partner Roebuck, to whom he wrote, "I sincerely wish you joy of this successful result, and hope it will make some return for the obligations I owe you." The visit of congratulation paid to his partner Roebuck, was delightful. Now were all their griefs "in the deep bosom of the ocean buried" by this recent success. Already they saw fortunes in their hands, so brightly shone the sun these few but happy days. But the old song has its lesson:

"I've seen the morning the gay hills adorning, I've seen it storming before the close of day."

Instead of instant success, trying days and years were still before them. A patent was decided upon, a matter of course and almost of formality in our day, but far from this at that time, when it was considered monopolistic and was highly unpopular on that account. Watt went to Berwick-on-Tweed to make the required declaration before a Master in Chancery. In August, 1768, we find him in London about the patent, where he became so utterly wearied with the delays, and so provoked with the enormous fees required to protect the invention, that he wrote his wife in a most despairing mood. She administered the right medicine in reply, "I beg you will not make yourself uneasy though things do not succeed as you wish. If the engine will not do, something else will; never despair." Happy man whose wife is his best doctor. From the very summit of elation, to which he had been raised by the success of the model, Watt was suddenly cast down into the valley of despair to find that only half of his heavy task was done, and the hill of difficulty still loomed before. Reaction took place, and the fine brain, so long strained to utmost tension, refused at intervals to work at high pressure. He became subject to recurring fits of despondency, aggravated, if not primarily caused by anxiety for his family, who could not be maintained unless he engaged in work yielding prompt returns.

We may here mention one of his lifelong traits, which revealed itself at times. Watt was no man of affairs. Business was distasteful to him. As he once wrote his partner, Boulton, he "would rather face a loaded cannon than settle a disputed account or make a bargain." Monetary matters were his special aversion. For any other form of annoyance, danger or responsibility, he had the lion heart. Pecuniary responsibility was his bogey of the dark closet. He writes that, "Solomon said that in the increase of knowledge there is increase of sorrow: if he had substituted *business* for knowledge it would have been perfectly true."

Roebuck shines out brilliantly in this emergency. He was always sanguine, and encouraged Watt to go forward. October, 1768, he writes:

You are now letting the most active part of your life insensibly glide away. A day, a moment, ought not to be lost. And you should not suffer your thoughts to be diverted by any other object, or even improvement of this [model], but only the speediest and most effectual manner of executing an engine of a proper size, according to your present ideas.

Watt wrote Dr. Small in January, 1769, "I have much contrived and little executed. How much would good health and spirits be worth to me!" and a month later, "I am still plagued with headaches and sometimes heartaches." Sleepless nights now came upon him. All this time, however, he was absorbed in his one engrossing task. Leupold's "Theatrim Machinarum," which fell into his hands, gave an account of the machinery, furnaces and methods of mine-working in the upper Hartz. Alas! The book was in German, and he could not understand it. He promptly resolved to master the language, sought out a Swiss-German dyer then settled in Glasgow whom he engaged to give him lessons. So German and the German book were both mastered. Not bad work this from one in the depths of despair. It has been before noted that for the same end he had successfully mastered French and Italian. So in sickness as in health his demon steam pursued him, giving him no rest.

Watt had a hard piece of work in preparing his first patent-specification, which was all-important in those early days of patent "monopolies" as these were considered. Their validity often turned upon a word or two too much or too little. It was as dangerous to omit as to admit. Professionals agree in opinion that Watt here displayed extraordinary ability.

In nothing has public opinion more completely changed than in its attitude toward patents. In Watt's day, the inventor who applied for a patent was a would-be monopolist. The courts shared the popular belief. Lord Brougham vehemently remonstrated against this, declaring that the inventor was entitled to remuneration. Every point was construed against the unfortunate benefactor, as if he were a public enemy attempting to rob his fellows. To-day the inventor is hailed as the foremost of benefactors.

Notable indeed is it that on the very day Watt obtained his first patent, January 5th, 1769, Arkwright got his spinning-frame patent. Only the year before Hargreaves obtained his patent for the spinning-jenny. These are the two inventors, with Whitney, the American inventor of the cotton-gin, from whose brains came the development of the textile industry in which Britain still stands foremost. Fifty-six millions of spindles turn to-day in the little island—more than all the rest of the civilised world can boast. Much later came Stephenson with his locomotive. Here is a record for a quartette of manual laborers in the truest sense, actual wage-earners as mechanics—Watt, Stephenson, Arkwright, and Hargreaves! Where is that quartette to be equalled?

Workingmen of our day should ponder over this, and take to heart the truth that manual mechanical labor is the likeliest career to develop mechanical inventors and lead them to such distinction as these benefactors of man achieved. If disposed to mourn the lack of opportunity, they should think of these working-men, whose advantages were small compared to those of our day.

The greatest invention of all, the condenser, is fully covered by the first patent of 1769. The best engine up to this time was the Newcomen, exclusively used for pumping water. As we have seen, it was an atmospheric engine, in no sense a steam engine. Steam was only used to force the heavy piston upward, no other work being done by it. All the pumping was done on the downward stroke. The condensation of the spent steam below the piston created a vacuum, which only facilitated the fall of the piston. This caused the cylinder to be cooled between each stroke and led to the wastage of about four-fifths of all the steam used. It was to save this that the condenser was invented, in obedience to Watt's law, as stated in his patent, that "the cylinder should be kept always as hot as the steam that entered it"; but it must be kept clearly in mind that Watt's "modified machines," under his first patent, only used steam to do work upon the upward stroke, where Newcomen used it only to force up the piston. The double-acting engine—doing work up and down—came later, and was protected in the second patent of 1780.

Watt knew better than any that although his model had been successful and was far beyond the Newcomen engine, it was obvious that it could be improved in many respects—not the least of his reasons for confidence in its final and more complete triumph.

To these possible improvements, he devoted himself for years. The records once again remind us that it was not one invention, but many, that his task involved. Smiles gives the following epitome of some of those pressing at this stage:

Various trials of pipe-condensers, plate-condensers and drum-condensers, steam-jackets to prevent waste of heat, many trials of new methods to tighten the piston band, condenser pumps, oil pumps, gauge pumps, exhausting cylinders, loading-valves, double cylinders, beams and cranks—all these contrivances and others had to be thought out and tested elaborately amidst many failures and disappointments.

There were many others.

All unaided, this supreme toiler thus slowly and painfully evolved the steam engine after long years of constant labor and anxiety, bringing to the task a union of qualities and of powers of head and hand which no other man of his time—may we not venture to say of all time—was ever known to possess or ever exhibited.

When a noble lord confessed to him admiration for his noble achievements, Watt replied, "The public only look at my success and not at the intermediate failures and uncouth constructions which have served me as so many steps to climb to the top of the ladder."

Quite true, but also quite right. The public have no time to linger over a man's mistakes. What concerns is his triumphs. We "rise upon our dead selves (failures) to higher things," and mistakes, recognised as such in after days, make for victory. The man who never makes mistakes never makes anything. The only point the wise man guards is not to make the same mistake twice; the first time never counts with the successful man. He both forgives and forgets that. One difference between the wise man and the foolish one!

It has been truly said that Watt seemed to have divined all the possibilities of steam. We have a notable instance of this in a letter of this period (March, 1769) to his friend, Professor Small, in which he anticipated Trevithick's use of high-pressure steam in the locomotive. Watt said:

I intend in many cases to employ the expansive force of steam to press on the piston, or whatever is used instead of one, in the same manner as the weight of the atmosphere is now employed in common fire engines. In some cases I intend to use both the condenser and this force of steam, so that the powers of these engines will as much exceed those pressed only by the air, as the expansive power of the steam is greater than the weight of the atmosphere. In other cases, when plenty of cold water cannot be had, I intend to work the engines by the force of steam only, and to discharge it into the air by proper outlets after it has done its office.

In these days patents could be very easily blocked, as Watt experienced with his improved crank motion. He proceeded therefore in great secrecy to erect the first large engine under his patent, after he had successfully made a very small one for trial. An outhouse near one of Dr. Roebuck's pits was selected as away from prying eyes. The parts for the new engine were partly supplied from Watt's own works in Glasgow and partly from the Carron works. Here the old trouble, lack of competent mechanics, was again met with. On his return from necessary absences, the men were usually found in face of the unexpected and wondering what to do next. As the engine neared completion, Watt's anxiety "for his approaching doom," he writes, kept him from sleep, his fears being equal to his hopes. He was especially sensitive and discouraged by unforeseen expenditure, while his sanguine partner, Roebuck, on the contrary, continued hopeful and energetic, and often rallied his pessimistic partner on his propensity to look upon the dark side. He was one of those who adhered to the axiom, "Never bid the devil good-morning till you meet him." Smiles believes that it is probable that without Roebuck's support Watt could never have gone on, but that may well be doubted. His anxieties probably found a needed vent in their expression, and left the indomitable do-or-die spirit in all its power. Watt's brain, working at high pressure, needed a safety valve. Mrs. Roebuck, wife-like, very properly entertained the usual opinion of devoted wives, that her husband was really the essential man upon whom the work devolved, and, that without him nothing could have been accomplished. Smiles probably founded his remark upon her words to Robison: "Jamie (Watt) is a queer lad, and, without the Doctor (her husband), his invention would have been lost. He won't let it perish." The writer knows of a business organisation in which fond wives of the partners were all full of dear Mrs. Roebuck's opinion. At one time, according to them, the sole responsibility rested upon three of four of these marvellous husbands, and never did any of the confiding consorts ever have reason to feel that their friend did not share to the fullest extent the highly praiseworthy opinion formed of his partners by their loving wives. The rising smile was charitably suppressed. In extreme cases a suggested excursion to Europe at the company's expense, to relieve Chester from the cruel strain, and enable him to receive the benefit of a wife's care and ever needful advice, was remarkably effective, the wife's fears that Chester's absence would prove ruinous to the business being overcome at last, though with difficulty.

Due allowance must be made for Mrs. Roebuck's view of the situation. There can be no doubt whatever, that Mr. Roebuck's influence, hopefulness and courage were of inestimable value at this period to the over-wrought and anxious inventor. Watt was not made of malleable stuff, and, besides, he was tied to his mission. He was bound to obey his genius.

The monster new engine, upon which so much depended, was ready for trial at last in September, 1769. About six months had been spent in its construction.Its success was indifferent. Watt had declared it to be a "clumsy job." The new pipe-condenser did not work well, the cylinder was almost useless, having been badly cast, and the old difficulty in keeping the piston-packing tight remained. Many things were tried for packing—cork, oiled rags, old hats (felt probably), paper, horse dung, etc., etc. Still the steam escaped, even after a thorough overhauling. The second experiment also failed. So great is the gap between the small toy model and the practical work-performing giant, a rock upon which many sanguine theoretical inventors have been wrecked! Had Watt been one of that class, he could never have succeeded. Here we have another proof of the soundness of the contention that Watt, the mechanic, was almost as important as Watt the inventor.

Watt remained as certain as ever of the soundness of his inventions. Nothing could shake his belief that he had discovered the true scientific mode of utilising steam. His failures lay in the impossibility of finding mechanics capable of accurate workmanship. There were none such at Carron, nor did he then know of any elsewhere.

Watt's letter to his friend, Dr. Small, at this juncture, is interesting. He writes:

You cannot conceive how mortified I am with this disappointment. It is a damned thing for a man to have his all hanging by asingle string. If I had wherewithal to pay the loss, I don't think I should so much fear a failure; but I cannot bear the thought of other people becoming losers by my schemes; and I have the happy disposition of always painting the worst.

Watt's timidity and fear of money matters generally have been already noted. He had the Scotch peasant's horror of debt—anything but that. This probably arises from the fact that the trifling sums owing by the poor to their poor neighbors who have kindly helped them in distress are actually needed by these generous friends for comfortable existence. The loss is serious, and this cuts deeply into grateful hearts. The millionaire's downfall, with large sums owing to banks, rich money-lenders, and wealthy manufacturers, really amounts to little. No one actually suffers, since imprisonment for debt no longer exists; hence "debt" means little to the great operator, who neither suffers want himself by failure nor entails it upon others.

To Watt, pressing pecuniary cares were never absent, and debt added to these made him the most afflicted of men. Besides this, he says, he had been cheated and was "unlucky enough to know." Wise man! Ignorance in such cases is indeed bliss. We should almost be content to be cheated as long as we do not find it out.

It was at such a crisis as this that another cloud, and a dark one, came. The sanguine, enterprising, kindly Roebuck was in financial straits. His pits had been much troubled by water, which no existing machinery could pump out. He had hoped that the new engine would prove successful and sufficiently powerful in time to avert the drowning of the pits, but this hope had failed. His embarrassments were so pressing that he was unable to pay the cost of the engine patent, according to agreement, and Watt had to borrow the money for this from that never-failing friend, Professor Black. Long may his memory be gratefully remembered? Watt had the delightful qualities which attracted friends, and those of the highest and best character, but among them all, though more than one might have been willing, none were both able and willing to sustain him in days of trouble except the famous discoverer of latent heat. When we think of Watt, we picture him holding Black by the one hand and small by the other, repeating to them

"I think myself in nothing else so happyAs in a soul remembering my dear friends."

The patent was secured—so much to the good—but Watt had already spent too much time upon profitless work, at least more time than he could afford. His duty to provide for the frugal wants of his family became imperative. "I had," he said, "a wife and children, and I saw myself growing gray without having any settled way of providing for them." He turned again to surveying and prospered, for few such men as Watt were to be found in those days, or in any day. With a record of Watt's work as surveyor, engineer, councillor, etc., our readers need not be troubled in detail. It should, however, be recorded that the chief canal schemes in Scotland in this, the day of canals for internal commerce, preceding the day of railroads that was to come, were entrusted to Watt, who continued to act as engineer for the Monkland Canal. While Watt was acting as engineer for this (1770-72), Dr. Small wrote him that he and Boulton had been talking of moving canal boats by the steam engine on the high-pressure principle. In his reply, September 30, 1770, Watt asks, "Have you ever considered a spiral oar for that purpose, or are you for two wheels?" To make his meaning quite plain, he gives a rough sketch of the screw propeller, with four turns as used to-day.

Thus the idea of the screw propeller to be worked by his own improved engine was propounded by Watt one hundred and thirty-five years ago.

This is a remarkable letter, and a still more remarkable sketch, and adds another to the many true forecasts of future development made by this teeming brain.

Watt also made a survey of the Clyde, and reported upon its proposed deepening. His suggestions remained unacted upon for several years, when the workwas begun, and is not ended even in our day, of making a trout and salmon stream into one of the busiest, navigable highways of the world. This year further improvements have been decided upon, so that the monsters of our day, with 16,000-horse-power turbine engines, may be built near Glasgow. Watt also made surveys for a canal between Perth and Coupar Angus, for the well-known Crinan Canal and other projects in the Western Highlands, as also for the great Caledonian and the Forth and Clyde Canals.

The Perth Canal was forty miles long through a rough country, and took forty-three days, for which Watt's fee, including expenses, was $400. Labor, even of the highest kind, was cheap in those times. We note his getting thirty-seven dollars for plans of a bridge over the Clyde. Watt prepared plans for docks and piers at Port Glasgow and for a new harbor at Ayr. His last and most important engineering work in Scotland was the survey of the Caledonian Canal, made in the autumn of 1773, through a district then without roads. "An incessant rain kept me," he writes, "for three days as wet as water could make me. I could scarcely preserve my journal book."

Suffice it to note that he saved enough money to be able to write, "Supposing the engine to stand good for itself, I am able to pay all my debts and some littlething more, so that I hope in time to be on a par with the world."

We are now to make one of the saddest announcements saving dishonor that it falls to man to make. Watt's wife died in childbed in his absence. He was called home from surveying the Caledonian Canal. Upon arrival, he stands paralysed for a time at the door, unable to summon strength to enter the ruined home. At last the door opens and closes and we close our eyes upon the scene—no words here that would not be an offence. The rest is silence.

Watt tried to play the man, but he would have been less than man if the ruin of his home had not made him a changed man. The recovery of mental equipoise proved for a time quite beyond his power. He could do all that man could do, "who could do more is none." The light of his life had gone out.

CHAPTER V

BOULTON PARTNERSHIP

After Watt was restored to himself the first subject which we find attracting him was the misfortunes of Roebuck, whose affairs were now in the hands of his creditors. "My heart bleeds for him," says Watt, "but I can do nothing to help him. I have stuck by him, indeed, until I have hurt myself." Roebuck's affairs were far too vast to be affected by all that Watt had or could have borrowed. For the thousand pounds Watt had paid on Roebuck's account to secure the patent, he was still in debt to Black. This was subsequently paid, however, with interest, when Watt became prosperous.

We now bid farewell to Roebuck with genuine regret. He had proved himself a fine character throughout, just the kind of partner Watt needed. It was a great pity that he had to relinquish his interest in the patent, when, as we shall see, it would soon have saved him from bankruptcy and secured him a handsome competence. He must ever rank as one of the men almost indispensable to Watt in the development of his engine, and a dear, true friend.

The darkest hour comes before the dawn, and so it proved here. As Roebuck retired, there appeared a star of hope of the first magnitude, in no less a person than the celebrated Matthew Boulton of Birmingham, of whom we must say a few words by way of introduction to our readers, for in all the world there was not his equal as a partner for Watt, who was ever fortunate in his friends. Of course Watt was sure to have friends, for he was through and through the devoted friend himself, and won the hearts of those worth winning. "If you wish to make a friend, be one," is the sure recipe.

Boulton was not only obviously the right man but he came from the right place, for Birmingham was the headquarters of mechanical industry. At this time, 1776, there was at last a good road to London. As late as 1747 the coach was advertised to run there in two days only "if the roads permit."

If skilled mechanics, Watt's greatest need, were to be found anywhere, it was here in the centre of mechanical skill, and especially was it in the celebrated works of Boulton, which had been bequeathed from worthy sire to worthy son, to be largely extended and more than ever preëminent.

Boulton left school early to engage in his father's business. When only seventeen years old, he had made several improvements in the manufacture of buttons, watch chains, and various trinkets, and had invented the inlaid steel buckles, which became so fashionable. It is stated that in that early day it was found necessary to export them in large quantities to France to be returned and sold in Britain as the latest productions of French skill and taste. It is well to get a glimpse of human nature as seen here. Fashion decides for a time with supreme indifference to quality. It is a question of the name.

At his father's death, the son inherited the business. Great credit belongs to him for unceasingly laboring to improve the quality of his products and especially to raise the artistic standard, then so low as to have already caused "Brummagem" to become a term of reproach. He not only selected the cleverest artisans, but he employed the best artists, Flaxman being one, to design the artistic articles produced. The natural result followed. Boulton's work soon gained high reputation. New and larger factories became necessary, and the celebrated Soho works arose in 1762. The spirit in which Boulton pursued business is revealed in a letter to his partner at Soho from London. "The prejudice that Birmingham hath so justly established against itself makes every fault conspicuous in all articles that have the least pretensions to taste." It may interest American readers familiar with One Dollar watches, rendered possible by production upon a large scale, that it was one of Boulton's leading ideas in that early day that articles in common use could be produced much better and cheaper "if manufactured by the help of the best machinery upon a large scale, and this could be successfully done in the making of clocks and timepieces." He promptly erected the machinery and started this new branch of business. Both King and Queen received him cordially and became his patrons. Soho works soon became famous and one of the show places of the country; princes, philosophers, poets, authors and merchants from foreign lands visited them and were hospitably received by Boulton.

He was besieged with requests to take gentlemen apprentices into the works, hundreds of pounds sometimes being offered as premium, but he resolutely declined, preferring to employ boys whom he could train up as workmen. He replies to a gentleman applicant, "I have built and furnished a house for the reception of one class of apprentices—fatherless children, parish apprentices, and hospital boys; and gentlemen's sons would probably find themselves out of place in such companionship."

It is not to be inferred that Boulton grew up an uncultured man because he left school very early. On the contrary, he steadily educated himself, devoting much time to study, so that with his good looks, handsome presence, the manners of the gentleman born, and knowledge much beyond the average of that class, he had little difficulty in winning for his wife a lady of such position in the county as led to some opposition on the part of members of her family to the suitor, but only "on account of his being in trade." There exists no survival of this objection in these days of American alliances with heirs of the highest British titles. We seem now to have as its substitute the condition that the father of the bride must be in trade and that heavily and to some purpose.

Boulton, like most busy men, had time, and an open mind, for new ideas. None at this time interested him so deeply as that of the steam engine. Want of water-power proved a serious difficulty at Soho. He wrote to a friend, "The enormous expense of the horse-power" (it was also irregular and sometimes failed) "put me upon thinking of turning the mill by fire. I made many fruitless experiments on the subject."

Boulton wrote Franklin, February 22, 1766, in London, about this, and sent a model he had made. Franklin replies a month later, apologising for the delay on account of "the hurry and anxiety I have been engaged in with our American affairs."

Tamer of lightning and tamer of steam, Franklin and Watt—one of the new, the other of the old branch of our English-speaking race—co-operating in enlarging the powers of man and pushing forward the chariot of progress—fit subject, this, for the sculptor and painter!

How much further the steam engine is to be the hand-maid of electricity cannot be told, for it seems impossible to set limits to the future conquests of the latter, which is probably destined to perform miracles un-dreamt of to-day, perhaps coupled in some unthought-of way, with radium, the youngest sprite of the weird, uncanny tribe of mysterious agents. Uranium, the supposed basis of the latest discovery, Radium, has only one-millionth part of the heat of the latter. The slow-moving earth takes twenty-four hours to turn upon its axis. Radium covers an equal distance while we pronounce its name. One and one-quarter seconds, and twenty-five thousand miles are traversed. Puck promises to put his "girdle round the earth in forty minutes." Radium would pass the fairy girdlist in the spin round sixteen hundred times. Thus truth, as it is being evolved in our day, becomes stranger than the wildest imaginings of fiction. Our century seems on the threshold of discoveries and advances, not less revolutionary, perhaps more so, than those that have sprung from steam and electricity. "Canst thou send lightnings to say 'Lo, here I am'?" silenced man. It was so obviously beyond his power until last century. Now he smiles as he reads the question. Is Tyndal's prophecy to be verified that "the potency of all things is yet to be found in matter"?

We may be sure the searching, restless brains of Franklin and Watt would have been meditating upon strange things these days if they were now alive.

Boulton is entitled to rank, so far as the writer knows, as the first man in the world worthy to wear Carlyle's now somewhat familiar title, "Captain of Industry" for he was in his day foremost in the industrial field, and before that, industrial organisations had not developed far enough to create or require captains, in Carlyle's sense.

Roebuck, while Watt's partner, was one of Boulton's correspondents, and told him of Watt's progress with the model engine which proved so successful. Boulton was deeply interested, and expressed a desire that Watt should visit him at Soho. This he did, on his return from a visit to London concerning the patent. Boulton was not at home, but his intimate friend, Dr. Small, then residing at Birmingham, a scientist and philosopher, whom Franklin had recommended to Boulton, took Watt in charge. Watt was amazed at what he saw, for this was his first meeting with trained and skilled mechanics, the lack of whom had made his life miserable. The precision of both tools and workmen sank deep. Upon a subsequent visit, he met the captain himself, his future partner, and of course, as like draws to like, they drew to each other, a case of mutual liking at first sight. We meet one stranger, and stranger he remains to the end of the chapter. We meet another, and ere we part he is a kindred soul. Magnetic attraction is sudden. So with these two, who, by a kind of free-masonry, knew that each had met his affinity. The Watt engine was exhaustively canvassed and its inventor was delighted that the great, sagacious, prudent and practical manufacturer should predict its success as he did. Shortly after this, Professor Robison visited Soho, which was a magnet that attracted the scientists in those days. Boulton told him that he had stopped work upon his proposed pumping engine. "I would necessarily avail myself of what I learned from Mr. Watt's conversation, and this would not be right without his consent."

It is such a delicate sense of honor as is here displayed that marks the man, and finally makes his influence over others commanding in business. It is not sharp practice and smart bargaining that tell. On the contrary, there is no occupation in which not only fair but liberal dealing brings greater reward. The best bargain is that good for both parties. Boulton and Watt were friends. That much was settled. They had business transactions later, for we find Watt sending a package containing "one dozen German flutes" (made of course by him in Glasgow), "at 5s. each, and a copper digester, £1:10." Boulton's people probably wished samples.

Much correspondence followed between Dr. Small and Watt, the latter constantly expressing the wish that Mr. Boulton could be induced to become partner with himself and Roebuck in his patents. Naturally the sagacious manufacturer was disinclined to associate himself with Mr. Roebuck, then in financial straits, but the position changed when he had become bankrupt and affairs were in the hands of creditors. Watt therefore renewed the subject and agreed to go and settle in Birmingham, as he had been urged to do. Roebuck's pitiable condition he keenly felt, and had done everything possible to ameliorate.

What little I can do for him is purchased by denying myself the conveniences of life my station requires, or by remaining in debt, which it galls me to the bone to owe. I shall be content to hold a very small share in the partnership, or none at all, provided I am to be freed from my pecuniary obligations to Roebuck and have any kind of recompense for even a part of the anxiety and ruin it has involved me in.

Thus wrote Watt to his friend Small, August 30, 1772. Small's reply pointed out one difficulty which deserves notice and commendation. "It is impossible for Mr. Boulton and me, or any other honest man, to purchase, especially from two particular friends, what has no market price, and at a time when they might be inclined to part with the commodity at an under value." This is an objection which to stock-exchange standards may seem "not well taken," and far too fantastical for the speculative domain, and yet it is neither surprising nor unusual in the realms of genuine business, in which men are concerned with or creating only intrinsic values.

The result so ardently desired by Watt was reached in this unexpected fashion. It was found that in the ordinary course of business Roebuck owed Boulton a balance of $6,000. Boulton agreed to take the Roebuck interest in the Watt patent for the debt. As the creditors considered the patent interest worthless, they gladly accepted. As Watt said, "it was only paying one bad debt with another."

Boulton asked Watt to act as his attorney in the matter, which he did, writing Boulton that "the thing is now a shadow; 'tis merely ideal, and will cost time and money to realise it." This as late as March 29, 1773, after eight years of constant experimentation, with many failures and disappointments, since the discovery of the separate condenser in 1765, which was then hailed, and rightly so, as the one thing needed. It remained the right and only foundation upon which to develop the steam engine, but many minor obstacles intervened, requiring Watt's inventive and mechanical genius to overcome.

The transfer of Roebuck's two-third interest to Boulton afterward carried with it the formation of the celebrated firm of Boulton and Watt. The latter arranged his affairs as quickly as possible. He had only made $1,000 for a whole year spent in surveying, and part of that he gave to Roebuck in his necessity, "so that I can barely support myself and keep untouched the small sum I have allotted for my visit to you." (Watt to Small, July 25, 1773). This is pitiable indeed—Watt pressed for money to pay his way to Birmingham upon important business.

The trial engine was shipped from Kinneil to Soho and Watt arrived in May, 1774, in Birmingham. Here a new life opened before him, still enveloped in clouds, but we may please ourselves by believing that through these the wearied and harassed inventor did not fail to catch alluring visions of the sun. Let us hope he remembered the words of the beautiful hymn he had no doubt often sung in his youth:

"Ye fearful saints, fresh courage takeThe clouds ye so much dreadAre big with mercy, and shall breakWith blessings on your head."

Partnership requires not duplicates, but opposites—a union of different qualities. He who proves indispensable as a partner to one man might be wholly useless, or even injurious, to another. Generals Grant and Sherman needed very different chiefs of staff. One secret of Napoleon's success arose from his being free to make his own appointments, choosing the men who had the qualities which supplemented his and cured his own shortcomings, for every man has shortcomings. The universal genius who can manage all himself has yet to appear. Only one with the genius to recognise others of different genius and harness them to his own car can approach the "universal." It is a case of different but coöperating abilities, each part of the complicated machine fitting into its right place, and there performing its duty without jarring.

Never were two men more "supplementary" to each other than Boulton and Watt, and hence their success. One possessed in perfection the qualities the other lacked. Smiles sums this up so finely that we must quote him:

Different though their characters were in most respects, Boulton at once conceived a hearty liking for him. The one displayed in perfection precisely those qualities which the other wanted. Boulton was a man of ardent and generous temperament, bold and enterprising, undaunted by difficulty, and possessing an almost boundless capacity for work. He was a man of great tact, clear perception, and sound judgment. Moreover, he possessed that indispensable quality of perseverance, without which the best talents are of comparatively little avail in the conduct of important affairs. While Watt hated business, Boulton loved it. He had, indeed, a genius for business—a gift almost as rare as that for poetry, for art, or for war. He possessed a marvellous power of organisation. With a keen eye for details, he combined a comprehensive grasp of intellect. While his senses were so acute, that when sitting in his office at Soho he could detect the slightest stoppage or derangement in the machinery of that vast establishment, and send his message direct to the spot where it had occurred, his power of imagination was such as enabled him to look clearly along extensive lines of possible action in Europe, America, and the East. *For there is a poetic as well as a commonplace side to business; and the man of business genius lights up the humdrum routine of daily life by exploring the boundless region of possibility wherever it may lie open before him.*

This tells the whole story, and once again reminds us that without imagination and something of the romantic element, little great or valuable is to be done in any field. He "runs his business as if it were a romance," was said upon one occasion. The man who finds no element of romance in his occupation is to be pitied. We know how radically different Watt was in his nature to Boulton, whose judgment of men was said to be almost unerring. He recognised in Watt at their first interview, not only the original inventive genius, but the indefatigable, earnest, plodding and thorough mechanic of tenacious grip, and withal a fine, modest, true man, who hated bargaining and all business affairs, who cared nothing for wealth beyond a very modest provision for old age, and who was only happy if so situated that without anxiety for money to supply frugal wants, he could devote his life to the development of the steam engine. Thus auspiciously started the new firm.

But Boulton was more than a man of business, continues Smiles; he was a man of culture, and the friend of educated men. His hospitable mansion at Soho was the resort of persons eminent in art, in literature, and in science; and the love and admiration with which he inspired such men affords one of the best proofs of his own elevation of character. Among the most intimate of his friends and associates were Richard Lovell Edgeworth, a gentleman of fortune, enthusiastically devoted to his long-conceived design of moving land-carriages by steam; Captain Keir, an excellent practical chemist, a wit and a man of learning; Dr. Small, the accomplished physician, chemist and mechanist; Josiah Wedgwood, the practical philosopher and manufacturer, founder of a new and important branch of skilled industry; Thomas Day, the ingenious author of "Sandford and Merton"; Dr. Darwin, the poet-physician; Dr. Withering, the botanist; besides others who afterward joined the Soho circle, not the least distinguished of whom were Joseph Priestley and James Watt.

The first business in hand was the reconstruction of the engine brought from Kinneil, which upon trial performed much better than before, wholly on account of the better workmanship attainable at Soho; but there still recurs the unceasing complaint that runs throughout the long eight years of trial—lack of accurate tools and skilled workmen, the difference in accuracy between the blacksmith standard and that of the mathematical-instrument maker. Watt and Boulton alike agreed that the inventions were scientifically correct and needed only proper construction. In our day it is not easy to see the apparently insuperable difficulty of making anything to scale and perfectly accurate, but we forget what the world of Watt was and how far we have advanced since.

Watt wrote to his father at Greenock, November, 1774: "The business I am here about has turned out rather successful; that is to say, the fire-engine I have invented is now going, and answers much better than any other that has yet been made." This is as is usual with the Scotch in speech, in a low key and extremely modest, on a par with the verdict rendered by the Dunfermline critic who had ventured to attend "the playhouse" in Edinburgh to see Garrick in Hamlet—"no bad." The truth was that, so pronounced were the results of proper workmanship, coupled with some of those improvements which Watt was constantly devising, the engine was so satisfactory as to set both Boulton and Watt to thinking about the patent which protected the invention. Six of the fourteen years for which it was granted had already passed. Some years would still be needed to ensure its general use, and it was feared that before the patent expired little return might be received. Much interest was aroused by the successful trial. Enquiries began to pour in for pumping engines for mines. The Newcomen had proved inadequate to work the mines as they became deeper, and many were being abandoned in consequence. The necessity for a new power had set many ingenious men to work besides Watt, and some of these were trying to adopt Watt's principles while avoiding his patent. Hatley, one of Watt's workmen upon the trial engine at the Carron works, had stolen and sold the drawings.

All this put Boulton and Watt on their guard, and the former hesitated to build the new works intended for the manufacture of steam engines upon a large scale with improved machinery. An extension of the patent seemed essential, and to secure this Watt proceeded to London and spent some time there, busy in his spare moments visiting the mathematical instrument shops of his youth, and attending to numerous commissions from Boulton. A second visit was paid to London, during which the sad intelligence of the death of his dear friend, Dr. Small, reached him. In the bitterness of his grief, Boulton writes him: "If there were not a few other objects yet remaining for me to settle my affections upon, I should wish also to take up my abode in the mansions of the dead." Watt's sympathetic reply reminds Boulton of the sentiments held by their departed friend—that, instead of indulging in unavailing sorrow, the best refuge is the more sedulous performance of duties. "Come, my dear sir," he writes, "and immerse yourself in this sea of business as soon as possible. Pay a proper respect to your friend by obeying his precepts. No endeavour of mine shall be wanting to make life agreeable to you."

Beautiful partnership this, not only of business, but also entering into the soul close and deep, comprehending all of life and all we know of death.

Professor Small, born 1734, was a Scot, who went to Williamsburg University, Virginia, as Professor of mathematics and natural philosophy. Thomas Jefferson was among his pupils. His health suffered, and he returned to the old home. Franklin introduced him to Boulton, writing (May 22, 1765):

I beg leave to introduce my friend Doctor Small to your acquaintance, and to recommend him to your civilities. I would not take this freedom if I were not sure it would be agreeable to you; and that you will thank me for adding to the number of those who from their knowledge of you must respect you, one who is both an ingenious philosopher and a most worthy, honest man. If anything new in magnetism or electricity, or any other branch of natural knowledge, has occurred to your fruitful genius since I last had the pleasure of seeing you, you will by communicating it greatly oblige me.

This man must have been one of the finest characters revealed in Watt's life. Altho he left little behind him to ensure permanent remembrance, the extraordinary tributes paid his memory by friends establish his right to high rank among the coterie of eminent men who surrounded Watt and Boulton. Boulton records that "there being nothing which I wish to fix in my mind so permanently as the remembrance of my dear departed friend, I did not delay to erect a memorial in the prettiest but most obscure part of my garden, from which you see the church in which he was interred." Dr. Darwin contributed the verses inscribed. Upon hearing of Small's illness Day hastened from Brussels to be present at the last hour.

Keir writes, announcing Small's death to his brother, the Rev. Robert Small, in Dundee, "It is needless to say how universally he is lamented; for no man ever enjoyed or deserved more the esteem of mankind. We loved him with the tenderest affection and shall ever revere his memory."

Watt's voluminous correspondence with Professor Small, previous to his partnership with Boulton, proves Small at that time to have been his intimate friend and counsellor. We scarcely know in all literature of a closer union between two men. Many verses of Tennyson's Memorial to Hallam could be appropriately applied to their friendship. Watt did not apparently give way to lamentations as Boulton and others did who were present at Small's death, probably because the receipt of Boulton's heart-breaking letter impressed Watt with the need of assuming the part of comforter to his partner, who was face to face with death, and had to bear the direct blow. Watt's tribute to his dear friend came later.

Future operations necessarily depended upon the extension of the patent. Boulton, of course, could not proceed with the works. There was as yet no agreement between Watt and Boulton beyond joint ownership in the patent. At this time, Watt's most intimate friend of youthful years in Glasgow University, Professor Robison, was Professor of mathematics in the Government Naval School, Kronstadt. He secured for Watt an appointment at $5,000 per annum, a fortune to the poor inventor; but although this would have relieved him from dependence upon Boulton, and meant future affluence, he declined, alleging that "Boulton's favours were so gracefully conferred that dependence on him was not felt." He made Watt feel "that the obligation was entirely upon the side of the giver." Truly we must canonise Boulton. He was not only the first "Captain of Industry," but also a model for all others to follow.

The bill extending the patent was introduced in Parliament February, 1775. Opposition soon developed. The mining interest was in serious trouble owing to the deepening of the mines and the unbearable expense of pumping the water. They had looked forward to the Watt engine soon to be free of patent rights to relieve them. "No monopoly," was their cry, nor were they without strong support, for Edmund Burke pleaded the cause of his mining constituents near Bristol.]

We need not follow the discussion that ensued upon the propriety of granting the patent extension. Suffice to say it was finally granted for a term of twenty-four years, and the path was clear at last. Britain was to have probably for the first time great works and new tools specially designed for a specialty to be produced upon a large scale. Boulton had arranged to pay Roebuck $5,000 out of the first profits from the patent in addition to the $6,000 of debt cancelled. He now anticipated payment of the thousand, at the urgent request of Roebuck's assignees, giving in so doing pretty good evidence of his faith in prompt returns from the engines, for which orders came pouring in. New mechanical facilities followed, as well as a supply of skilled mechanics.

The celebrated Wilkinson now appears upon the scene, first builder of iron boats, and a leading iron-founder of his day, an original Captain of Industry of the embryonic type, who began working in a forge for three dollars a week. He cast a cylinder eighteen inches in diameter, and invented a boring machine which bored it accurately, thus remedying one of Watt's principal difficulties. This cylinder was substituted for the tin-lined cylinder of the triumphant Kinneil engine. Satisfactory as were the results of the engine before, the new cylinder improved upon these greatly. Thus Wilkinson was pioneer in iron ships, and also in ordering the first engine built at Soho—truly an enterprising man. Great pains were taken by Watt that this should be perfect, as so much depended upon a successful start. Many concerns suspended work upon Newcomen engines, countermanded orders, or refrained from placing them, awaiting anxiously the performance of this heralded wonder, the Watt engine. As it approached completion, Watt became impatient to test its powers, but the prudent, calm Boulton insisted that not one stroke be made until every possible hindrance to successful working had been removed. He adds, "Then, in the name of God, fall to and do your best." Admirable order of battle! It was "Be sure you're right, then go ahead," in the vernacular. Watt acted upon this, and when the trial came the engines worked "to the admiration of all." The news of this spread rapidly. Enquiries and orders for engines began to flow in. No wonder when we read that of thirty engines of former makers in one coal-mining district only eighteen were at work. The others had failed. Boulton wrote Watt to tell Wilkinson to get a dozen cylinders cast and bored ... I have fixed my mind upon making from twelve to fifteen reciprocating engines and fifty rotative engines per annum. Of all the toys and trinkets we manufacture at Soho, none shall take the place of fire-engines in respect of my attention.

The captain was on deck, evidently. Sixty-five engines per year—prodigious for these days—nothing like this was ever heard of before. Two thousand per year is the record of one firm in Philadelphia to-day, but let us boast not. Perhaps one hundred and twenty-nine years hence will have as great a contrast to show. The day of small factories, as of small nations, is past. Increasing magnitude, to which it is hard to set a limit, is the order of the day.

So far all was well, the heavy clouds that had so long hovered menacingly over Boulton and Watt had been displaced once more by clear skies. But no new machinery or new manufacturing business starts without accidents, delays and unexpected difficulties. There was necessarily a long period of trial and disappointment for which the sanguine partners were not prepared. As before, the chief trouble lay in the lack of skilled workmen, for although the few original men in Soho were remarkably efficient, the increased demand for engines had compelled the employment of many new hands, and the work they could perform was sadly defective. Till this time, it is to be remembered there had been neither slide lathes, planing machines, boring tools, nor any of the many other devices which now ensure accuracy. All depended upon the mechanics' eye and hand, if mechanics they could be called. Most of the new hands were inexpert and much given to drink. Specialisation had to be resorted to—one thing for each workman, in the fashioning of which practice made perfect. This system was introduced with success, but the training of the men took time. Meanwhile work already turned out and that in progress was not up to standard, and this caused infinite trouble. One very important engine was "The Bow" for London, which was shipped in September. The best of the experts, Joseph Harrison, was sent to superintend its erection. Verbal instructions Watt would not depend upon; Harrison was supplied in writing with detailed particulars covering every possible contingency. Constant communication between them was kept up by letter, for the engine did not work satisfactorily, and finally Watt himself proceeded to London in November and succeeded in overcoming the defects. Harrison's anxieties disabled him, and Boulton wrote to Dr. Fordyce, a celebrated doctor of that day, telling him to take good care of Harrison, "let the expense be what it will." Watt writes Boulton that Harrison must not leave London, as "a relapse of the engine would ruin our reputation here and elsewhere." The Bow engine had a relapse, however, which happened in this way. Smeaton, then the greatest of the engineers, requested Boulton's London agent to take him to see the new engine. He carefully examined it, called it a "very pretty engine," but thought it too complicated a piece of machinery for practical use. There was apparently much to be said for this opinion, for we clearly see that Watt was far in advance of his day in mechanical requirements. Hence his serious difficulties in the construction of the complex engine, and in finding men capable of doing the delicately accurate work which was absolutely indispensable for successful working.

Before leaving, Smeaton made the engineer a gift of money, which he spent in drink. The drunken engineman let the engine run wild, and it was thrown completely out of order. The valves—the part of the complicated machine that required the most careful treatment—were broken. He was dismissed, and, repairs being made, the engine worked satisfactorily at last. In Watt's life, we meet drunkenness often as a curse of the time. We have the satisfaction of knowing that our day is much freer from it. We have certainly advanced in the cure of this evil, for our working-men may now be regarded as on the whole a steady sober class, especially in America, where intemperance has not to be reckoned with.

We see the difference between the reconstructed Kinneil engines where Boulton's "mathematical instrument maker's" standard of workmanship was possible "because his few trained men capable of such work were employed." The Kinneil engine, complicated as it was in its parts, being thus accurately reconstructed, did the work expected and more. The Bow engines and some others of the later period, constructed by ordinary workmen capable only of the "blacksmith's" standard of finish, proved sources of infinite trouble.

Watt had several cases of this kind to engross his attention, all traceable to the one root, lack of the skilled, sober workmen, and the tools of precision which his complex (for his day, very complex) steam engine required. The truth is that Watt's engine in one sense was born before its time. Our class of instrument-making mechanics and several new tools should have preceded it; then, the science of the invention being sound, its construction would have been easy. The partners continued working in the right direction and in the right way to create these needful additions and were finally successful, but they found that success brought another source of annoyance. Escaping Scylla they struck Charybdis. So high did the reputation of their chief workmen rise, that they were early sought after and tempted to leave their positions. Even the two trained fitters sent to London to cure the Bow engine we have just spoken of were offered strong inducements to take positions in Russia. Watt writes Boulton, May 3, 1777, that he had just heard a great secret to the effect that Carless and Webb were probably going beyond sea, $5,000 per year having been offered for six years. They were promptly ordered home to Soho and warrants obtained for those who had attempted to induce them to abscond (strange laws these days!), "even though Carless be a drunken and comparatively useless fellow." Consider Watt's task, compelled to attempt the production of his new engines, complicated beyond the highest existing standard, without proper tools and with such workmen as Carless, whom he was glad to get and determined to keep, drunken and useless as he was.

French agents appeared and tried to bribe some of the men to go to Paris and communicate Watt's plans to the contractor who had undertaken to pump water from the Seine for the supply of Paris. The German states sent emissaries for a similar purpose, and Baron Stein was specially ordered by his government to master the secret of the Watt engine, to obtain working plans, and bring away workmen capable of constructing it, the first step taken being to obtain access to the engine-rooms by bribing the workmen. All this is so positively stated by Smiles that we must assume that he quotes from authentic records. It is clear at all events that the attention of other nations was keenly drawn to the advent of an agency that promised to revolutionise existing conditions. Watt himself, at a critical part of his career (1773), as we have seen, had been tempted to accept an offer to enter the imperial service of Russia, carrying the then munificent salary of $5,000 per annum. Boulton wrote him: "You're going to Russia staggers me.... I wish to advise you for the best without regard to self, but I find I love myself so well that I should be very sorry to have you go, and I begin to repent sounding your trumpet at the Ambassador's."

The imperial family of Russia were then much interested in the Soho works. The empress stayed for some time at Boulton's house, "and a charming woman she is," writes her host. Here is a glimpse of imperial activity and wise attention to what was going on in other lands which it was most desirous to transplant to their own. The emperor, and no less his wife, evidently kept their eyes open during their travels abroad. Imperial progresses we fear are seldom devoted to such practical ends, although the present king of Britain and his nephew the German emperor would not be blind to such things. It is a strange coincidence that the successor of this emperor, Tsar Nicholas, when grand duke, should have been denied admission to Soho works. Not that he was personally objected to, but that certain people of his suite might not be disinclined to take advantage of any new processes discovered. So jealously were improvements guarded in these days?

Another source of care to the troubled Watt lay here. Naturally, only a few such men had been developed as could be entrusted to go to distant parts in charge of fellow-workmen and erect the finished engines. A union of many qualities was necessary here. Managers of erection had to be managers of men, by far the most complicated and delicate of all machinery, exceeding even the Watt engine in complexity. When the rare man was revealed, and the engine under his direction had proved itself the giant it was reputed, ensuring profitable return upon capital invested in works hitherto unproductive, as it often did, the sagacious owner would not readily consent to let the engineer leave. He could well afford to offer salary beyond the dreams of the worker, to a rider who knew his horse and to whom the horse took so kindly. The engineer loved *his* engine, the engine which *he* had seen grow in the shop under his direction and which *he* had wholly erected.

McAndrew's Song of Steam tells the story of the engineer's devotion to his engine, a song which only Kipling in our day could sing. The Scotch blood of the MacDonalds was needed for that gem; Kipling fortunately has it pure from his mother. McAndrew is homeward bound patting *his* mighty engine as she whirls, and crooning over his tale:

That minds me of our Viscount loon—Sir Kenneth's kin—the chapWi' Russia leather tennis-shoon an' spar-decked yachtin'-cap.I showed him round last week, o'er all—an' at the last says he:"Mister M'Andrew, don't you think steam spoils romance at sea?"Damned ijjit! I'd been doon that morn to see what ailed the throws, Manholin', on my back—the cranks three inches off my nose.Romance! Those first-class passengers they like it very well,Printed an' bound in little books; but why don't poets tell?I'm sick of all their quirks an' turns—the loves and doves they dream—Lord, send a man like Robbie Burns to sing the Song o' Steam!To match wi' Scotia's noblest speech yon orchestra sublime,Whaurto—uplifted like the Just—the tail-rods mark the time.The crank-throws give the double-bass, the feed-pump sobs an' heaves,An' now the main eccentrics start their quarrel on the sheaves:Her time, her own appointed time, the rocking link-head bides,Till—hear that note?—the rod's return whings glimmerin' through the guides.They're all awa'! True beat, full power, the clangin' chorus goesClear to the tunnel where they sit, my purrin' dynamos.Interdependence absolute, foreseen, ordained, decreed,To work, ye'll note, at any tilt an' every rate o' speed.Fra' skylight lift to furnace-bars, backed, bolted, braced an' stayed,An' singin' like the Mornin' Stars for joy that they are made;While, out o' touch o' vanity, the sweatin' thrust-block says:"Not unto us the praise, oh man, not unto us the praise!"Now, a' together, hear them lift their lesson—theirs an' mine:"Law, Order, Duty an' Restraint, Obedience, Discipline!"Mill, forge an' try-pit taught them that when roarin' they arose,An' whiles I wonder if a soul was gied them wi' the blows.Oh for a man to weld it then, in one trip-hammer strain,Till even first-class passengers could tell the meanin' plain!But no one cares except mysel' that serve an' understandMy seven-thousand horse-power here. Eh, Lord! They're grand—they're grand! Uplift am I? When first in store the new-made beasties stood,Were ye cast down that breathed the Word declarin' all things good?Not so! O' that world-liftin' joy no after-fall could vex, ye've left a glimmer still to cheer the Man—the Artifex! *That* holds, in spite o' knock and scale, o' friction, waste an' slip,An' by that light—now, mark my word—we'll build the Perfect Ship.I'll never last to judge her lines or take her curve—not I.But I ha' lived and I ha' worked. Be thanks to Thee, Most High!

So the McAndrews of Watt's day were loth to part from *their* engines, this feeling being in the blood of true engineers. On the other hand, just such men, in numbers far beyond the supply, were needed by the builders, who in one sense were almost if not quite as deeply concerned as the owners, in having proved, capable, engine managers remain in charge of their engines, thus enhancing their reputation. Endless trouble ensued from the lack of managing enginemen, a class which had yet to be developed, but which was sure to arise in time through the educative policy adopted, which was already indeed slowly producing fruit.

Meanwhile, to meet the present situation, Watt resolved to simplify the engine, taking a step backward, which gives foundation for Smeaton's acute criticism upon its complexity. We have seen that the working of steam expansively was one of Watt's early inventions. Some of the new engines were made upon this plan, which involved the adoption of some of the most troublesome of the machinery. It was ultimately decided that to operate this was beyond the ability of the obtainable enginemen of the day.

It must not be understood that expansion was abandoned. On the contrary, it was again introduced by Watt at a later stage and in better form. Since his time it has extended far beyond what he could have ventured upon under the conditions of that day. "Yet," as Kelvin says, "the triple and quadruple expansion engine of our day all lies in the principle Watt had so fully developed in his day."

CHAPTER VI

REMOVAL TO BIRMINGHAM

Watt's permanent settlement in Birmingham had for some time been seen to be inevitable, all his time being needed there. Domestic matters, including the care of his two children, with which he had hitherto been burdened, pressed hard upon him, and he had been greatly depressed by finding his old father quite in his dotage, although he was not more than seventy-five. Watt was alone and very unhappy during a visit he made to Greenock.

Before returning to Birmingham, he married Miss MacGregor, daughter of a Glasgow man of affairs, who was the first in Britain to use chlorine for bleaching, the secret of which Berthollet, its inventor, had communicated to Watt.

Pending the marriage, it was advisable that the partnership with Boulton as hitherto agreed upon should be executed. Watt writes so to Boulton, and the arrangement between the partners is indicated by the following passage of Watt's letter to him:

As you may have possibly mislaid my missive to you concerning the contract, I beg just to mention what I remember of the terms.

1. I to assign to you two-thirds of the property of the invention.

2. You to pay all expenses of the Act or others incurred before June, 1775 (the date of the Act), and also the expense of future experiments, which money is to be sunk without interest by you, being the consideration you pay for your share.

3. You to advance stock-in-trade bearing interest, but having no claim on me for any part of that, further than my intromissions; the stock itself to be your security and property.

4. I to draw one-third of the profits so soon as any arise from the business, after paying the workmen's wages and goods furnished, but abstract from the stock-in-trade, excepting the interest thereof, which is to be deducted before a balance is struck.

5. I to make drawings, give directions, and make surveys, the company paying for the travelling expenses to either of us when upon engine business.

6. You to keep the books and balance them once a year.

7. A book to be kept wherein to be marked such transactions as are worthy of record, which, when signed by both, to have the force of the contract.

8. Neither of us to alienate our share of the other, and if either of us by death or otherwise shall be incapacitated from acting for ourselves, the other of us to be the sole manager without contradiction or interference of heirs, executors, assignees or others; but the books to be subject to their inspection, and the acting partner of us to be allowed a reasonable commission for extra trouble.

9. The contract to continue in force for twenty-five years, from the 1st of June, 1775, when the partnership commenced, notwithstanding the contract being of later date.

10. Our heirs, executors and assignees bound to observance.

11. In case of demise of both parties, our heirs, etc., to succeed in same manner, and if they all please, they may burn the contract.

If anything be very disagreeable in these terms, you will find me disposed to do everything reasonable for your satisfaction.

Boulton's reply was entirely satisfactory, and upon this basis the arrangement was closed.

Watt, with his usual want of confidence in himself in business affairs, was anxious that Boulton should come to him at Glasgow and arrange all pecuniary matters connected with the marriage. Watt had faced the daughter and conquered, but trembled at the thought of facing the father-in-law. He appeals to his partner as follows:

I am afraid that I shall otherwise make a very bad bargain in money matters, which wise men like you esteem the most essential part, and I myself, although I be an enamoured swain, do not altogether despise. You may perhaps think it odd that in the midst of my friends here I should call for your help; but the fact is that from several reasons I do not choose to place that confidence in any of my friends here that would be necessary in such a case, and I do not know any of them that have more to say with the gentleman in question than I have myself. Besides, you are the only person who can give him satisfactory information concerning my situation.

This being impracticable, as explained by Boulton, who thoroughly approved of the union, the partnership and Boulton's letter were accepted by the judicious father-in-law as satisfactory evidence that his daughter's future was secure. Boulton states in his letter, July, 1776:

It may be difficult to say what is the value of your property in partnership with me. However, I will give it a name, and I do say that I would willingly give you two, or perhaps three thousand pounds for your assignment of your third part of the Act of Parliament. But I should be sorry to make you so bad a bargain, or to make any bargain at all that tended to deprive me of your friendship, acquaintance, and assistance, hoping that we shall harmoniously live to wear out the twenty-five years, which I had rather do than gain a Nabob's fortune by being the sole proprietor.

This is the kind of expression from the heart to make a partner happy and resolve to do his utmost for one who in the recipient's heart had transposed positions, and is now friend first, and partner afterward.

The marriage took place in July, 1776. Two children were born, both of whom died in youth. Mrs. Watt lived until a ripe old age and enjoyed the fruits of her husband's success and fame. She died in 1832. Arago praises her, and says "Various talents, sound judgment, and strength of mind rendered her a worthy companion."

It is difficult to realise that many yet with us were contemporaries of Mrs. Watt, and not a few yet living were contemporaries of Watt himself, for he did not pass away until 1819, eighty-six years ago, so much a thing of yesterday is the material development and progress of the world, which had its basis, start and accomplishment in the steam engine.

The reasons given by Boulton for being unable to proceed to the side of his friend and partner in Glasgow, shed clear light upon the condition of affairs at Soho. Their London agent, like Watt, was also to be married and would be absent. Fothergill had to proceed to London. Scale, one of the managers, was absent. Important visitors were constantly arriving. Said Boulton:

Our copper bottom hath plagued us very much by steam leaks, and therefore I have had one cast (with its conducting pipe) all in one piece; since which the engine doth not take more than 10 feet of steam, and I hope to reduce that quantity, as we have just received the new piston, which shall be put in and at work tomorrow. Our Soho engine never was in such good order as at present. Bloomfield and Willey (engines) are both well, and I doubt not but Bow engine will be better than any of 'em.

He concludes, "I did not sleep last night, my mind being absorbed by steam." Means for increasing the heating surface swept through his mind, by applying "in copper spheres within the water," the present flue system, also for working steam expansively, "being clear the principle is sound."

To add to Boulton's anxieties, he had received a summons to attend the Solicitor-General next week in opposition to Gainsborough, a clergyman who claimed to be the original inventor. "This is a disagreeable circumstance, particularly at this season, when you are absent. Harrison is in London and idleness is in our engine shop."

Watt wrote Boulton on July 28, 1776, apologising for his long absence and stating he was now ready to return, and would start "Tuesday first" for Liverpool, where he expected to meet Boulton. Meanwhile, the latter had been called to London by the Gainsborough business. A note from him, however, reached Watt at Liverpool, in which he says, "As to your absence, say nothing about it. I will forgive it this time, *provided you promise me never to marry again.*"

In due time, Mr. and Mrs. Watt arrived and settled early in August, 1776, in Birmingham, which was hereafter to be their permanent home, although, as we shall see, Watt never ceased to keep in close touch with his native town of Greenock and his Glasgow friends. His heart still warmed to the tartan, the soft, broad Scotch accent never forsook him; nor, we may be sure, did the refrain ever leave his heart——

And may dishonour blot our nameAnd quench our household fires,If me or mine forget thy name,Thou dear land of my Sires,

Many a famous Scot has the fair South in recent times called to her—Stephenson, Ruskin, Carlyle, Mill, Gladstone and others—but never before or since, one whose work was the transformation of the world.

At last we have Watt permanently settled alongside the great works to which he was hereafter to devote his rare abilities until his retirement at the expiration of the partnership in 1800. His labors at Soho soon began to tell. The works increased their celebrity beyond all others then known, for materials, workmanship and invention.

The mines of Cornwall promised to become unworkable; indeed, many already had became so. The Newcomen engines could no longer drain the deepened mines. Several orders for Watt engines had been received, and as much depended upon the success of the first, Watt resolved to superintend its erection himself. Mrs. Watt and he started over the terrible road into Cornwall, and had to take up their abode with the superintendent of the mine, there being no other house for miles around. Naturally the builders and attendants of the Newcomen engine viewed Watt's invasion of their district with no kindly feelings. Great jealousy arose and Watt's sensitive nature was sorely tried. Many attempts to thwart him were met with, and, taken altogether, his life in Cornwall was far from agreeable.

The engine was erected, the day of trial came, mining men, engineers, mining proprietors and others assembled from all quarters to see the start. Many of the spectators interested in other engines would not have shed tears had it failed, but it started splendidly making eleven eight-foot strokes per minute, which broke the record. Three cheers for the Scotch engineer! It soon worked with greater power and more steadily, and "forked" more water than the ordinary engines with only about one-third the consumption of coal. Watt wrote:

I understand all the West Country captains are to be here tomorrow to see the prodigy. The velocity, violence, magnitude, and horrible noise of the engine give universal satisfaction to all beholders, believers or not. I have once or twice trimmed the engine to end the stroke gracefully and to make less noise, but Mr. Wilson cannot sleep without it seems quite furious, so I have left it to the enginemen; and, by the by, the noise seems to convey great ideas of its power to the ignorant, who seem to be no more taken with modest merit in an engine than in a man.

Well said, modest, reserved philosopher with vast horse-power in that big head of yours, working in the closet noiselessly, driving deep but silently into the bosom of nature's secrets, pumping her deepest mines, discovering and bringing to the surface the genius which lay in steam to do your bidding and revolutionise life on earth! In this, the first triumph, there was recompense for all the trials Watt and his wife had endured in Cornwall.

Readers will note that no workman had yet been developed who could be trusted to erect the engine. The master inventor had to go himself as the mechanical genius certain to cure all defects and ensure success. This shows how indispensable Watt was.

Orders now flowed in, and Watt was needed to prepare the plans and drawings, no one being capable of relieving him of this. To-day we have draftsmen by the thousand to whom it would be easy routine work, as we have thousands to whom the erection of the Watt engine would be play. Watt was everywhere. At length he had to confess that "a very little more of this hurrying and vexation would knock me up altogether." At this moment he had just been called to return to Cornwall to erect the second engine. He says "I fancy I must be cut in pieces and a portion sent to every tribe in Israel." We may picture him reciting in Falstaffian mood, "Would my name were not so terrible to the enemy (deep-mine water) as it is. There can't a drowned-out mine peep its head out but I'm thrust upon it. Well, well, it always was the trick of my countrymen to make a good thing too common. Better rust to death than be scoured to nothing by this perpetual motion."

Watt had a hard time of it in Cornwall during his next stay there, for he had to go again. He arrives at Redruth to find many troubles.

Forbes' eduction-pipe is a vile job, he writes, and full of holes. The cylinder they have cast for Chacewater is still worse, for it will hardly do at all. The Soho people have sent here Chacewater pipe instead of Wheal Union, and the gudgeon pipe has not arrived with the nozzles. These repeated disappointments will ruin our credit in the country, and I cannot stay here to bear the shame of such failures of promise.

It is easy for present-day captains of industry to plume themselves upon their ability to select men sure to succeed well with any undertaking, and assume that Watt lacked the indispensable talent for selection, but he had been driven by sad experience to trust none but himself, the skilled workmen needed to co-operate with him not yet having been developed.

We have not touched upon another source of great anxiety to him at this time. The enterprising Boulton would not have been the organiser he was unless blessed with a sanguine disposition and the capacity for shedding troubles. The business was rapidly extending in many branches, all needing capital; the engine business, promising though it was, was no exception. Little money was yet due from sales and much had been spent developing the invention.

Boulton's letter to Watt constantly urged cash collections, while mine-owners were not disposed to pay until further tests were made. Boulton suggested loans from Truro bankers on security of the engines, but Watt found this impracticable. The engines were doing astonishingly well to-day, but who could ensure their lasting qualities? Watt shows good judgment in suggesting that Wilkinson, the famous foundryman, should be taken into partnership. He urges his enterprising partner to apply the pruning knife and cut down expenses naively assuring him that "he was practising all the frugality in his power." As Watt's personal expenses then were only ten dollars per week, a smile rises at the prudent Scot's possible contribution to reduction in expenditure. But he was on the right lines, and at least gave Boulton the benefit of example. Watt was never disposed to look on the bright side of things, and to add to Boulton's load, the third partner, Fothergill, was even more desponding than Watt. When Boulton went away to raise means, he was pursued by letters from Fothergill telling him day by day of imperative needs. In one he was of opinion that "the creditors must be called together; better to face the worst than to go on in the neck-and-neck race with ruin." Boulton would hurry back to quiet Fothergill and keep the ship afloat. Here he shines out resplendently. He proved equal to the emergency. His courage and determination rose in proportion to the difficulties to be overcome, borne up by his invariable hope and unshakable belief in the value of Watt's condensing engine, he triumphed at last, pledging, as security for a loan of $70,000, the royalties derivable from the engine patents, and an annuity for a loan of $35,000 more. So small a sum as $105,000 sufficed to keep afloat the big ship laden with all their treasures.

There was a period of great depression in Britain when Boulton and Watt were thus in deep water, and at such times credit is sensitive in the extreme. A small balance on the right side performs wonders. This recalls to the writer how, once in the history of his own firm, credit was kept high during a panic by using the identical sum Boulton raised, $70,000, from a reserve fund that had been laid away and came in very opportunely at the critical time. Every single dollar weighs a hundredfold when credit trembles in the balance. A leading nerve specialist in New York once said that the worst malady he had to treat was the man of affairs whose credit was suspected. His unfailing remedy was: "Call your creditors together, explain all and ask their support. I can then do you some good, but not till then." His patients who did this found themselves restored to vigor. They were supported by creditors and all was bright once more. The wise doctor was sound in his advice. If the firm has neither speculated nor gambled (synonymous terms), nor lived extravagantly, nor endorsed for others, and the business is on a solid foundation, no people have so much at stake in sustaining it as the creditors; they will rally round it and think more of the firm than ever, because they will see behind their money the best of all securities—men at the helm who are not afraid and know how to meet a storm.

Boulton's timid partners no doubt were amazed that he was so blind to the dangers which they with clearer vision saw so clearly. How deluded they were. We may be sure neither of them saw the danger half as vividly as he, but it is not the part of a leader to reveal to his fellows all that he sees or fears. His part is to look dangers steadily in the face and challenge them. It is the great leader who inspires in his followers contempt for the danger which he sees in much truer proportion than they. This Boulton did, for behind all else in his character there lay the indomitable will, the do or die resolve. He had staked his life upon the hazard of a die and he would stand the cost. "But if we fail," often said the timid pair to him, as Macbeth did to his resolute partner, and the same answer came, "*We* fail." That's all. "One knockdown will not finish this fight. We'll get up again, never fear. We know no such word as fail."]

One source of serious trouble arose from Watt and Boulton having been so anxious at first to introduce their engines that they paid small regard to terms. When their success was proved, they offered to settle, taking one-third the value of the fuel saved. This was a liberal offer, for, in addition to the mine-owners saving two-thirds of the former cost of fuel consumed by the previous engines, mines became workable, which without the Watt engine must have been abandoned. These terms however were not accepted, and a long series of disputes arose, ending in some cases only with the patent-right itself. It was resolved that all future engines should be furnished only upon the terms before stated, Watt declaring that otherwise he would not put pen to paper to make new drawings. "Let our terms be moderate," he writes, "and, if possible, consolidated into money *a priori*, and it is certain we shall get *some* money, enough to keep us out of jail, in continual apprehension of which I live at present." Imprisonment for debt, let it be remembered, had not been abolished. One of the most beneficent forward steps that our time can boast of is the Bankruptcy Court. However hard we may yet be upon offenders against us, society, through humane laws, forgives our debtors in money matters, and gives a clear bill of health after honorable acquittal in bankruptcy, and a fresh start.

The result proved Watt's wisdom. His engines were needed to save the mines. No other could. Applications came in freely upon his terms, and as Watt was a poor hand at bargaining, he insisted that Boulton should come to Cornwall and attend to that part.

Meanwhile great attention was being paid to the works and all pertaining to the men and methods. The firm established perhaps the first benefit society of workmen. Every one was a member and contributed according to his earnings. Out of this fund payments were made to the sick or disabled in varying amounts. No member of the Soho Friendly Society, except a few irreclaimable drunkards, ever came upon the parish.

When Boulton's son came of age, seven hundred were dined. No well-behaved workman was ever turned adrift. Fathers employed introduced their sons into the works and brought them up under their own eye, watching over their conduct and mechanical training. Thus generation after generation followed each other at Soho works.

On another occasion Boulton writes Watt in Cornwall, "I have thought it but respectful to give our folks a dinner to-day. There were present Murdoch, Lawson, Pearson, Perkins, Malcom, Robert Muir, all Scotchmen, John Bull and Wilson and self, for the engines are now all finished and the men have behaved well and are attached to us."

Six Scotch and three English in the English works of Soho thought worthy of dining with their employer! It was, we may be sure, a very rare occurrence in that day, but worthy of the true captain of industry. Here is an early "invasion" from the north. We are reminded of Sir Charles Dilke's statement in his "Greater Britain," that, in his tour round the world, he found ten Scotchmen for every Englishman in high position. Owing, of course, to the absence of scope at home the Scot has had to seek his career abroad.

A master-stroke this, probably the first dinner of its kind in Britain, and no doubt more highly appreciated by the honored guests than an advance in wages. Splendid workmen do not live upon wages alone. Appreciation felt and shown by their employer, as in this case, is the coveted reward.

We have read how Watt was much troubled in Scotland with poor mechanics. Not one good craftsman could he then find. After seeing Soho, where the standard was much higher, he declared that the Scotch mechanic was very much inferior; he was prejudiced against them. Murdoch, however, the first

Scot at Soho, soon eclipsed all, and no doubt under his wing other Scots gained a trial with the result indicated. It is very significant that even in the earliest days of the steam engine, Scotchmen should exhibit such talent for its construction, forecasting their present pre-eminence in marine engineering.

Small wonder that the Soho works became the model for all others. The last words in Boulton's letter, "and are attached to us," tell the story. No danger of strikes, of lockouts, or quarrels of any kind in such establishments as that of Boulton and Watt, who proved that they in turn were attached to their men. Mutual attachment between employers and employed is the panacea for all troubles—yes, better than a panacea, the preventer of troubles.

After repeated calls from Watt, Boulton took the journey to Cornwall in October, 1778, although Fothergill was again uttering lamentable prophecies of impending ruin, and the London agent was imploring his presence there upon financial matters pressing in the extreme. Boulton succeeded in borrowing $10,000 from Truro bankers on the security of engines erected, and settled several disputes, getting $3,500 per year royalty for one engine and $2,000 per year for another. At last, after nine years of arduous labor since the invention was hailed as successful, the golden harvest so long expected began to replenish the empty treasury. The heavy liabilities, however, remained a source of constant anxiety. No remedy could be found against "this consumption of the purse."

Watt had again to encounter the lack of competent, sober workmen to run engines. The Highland blood led him at last into severe measures, and he insisted upon discharging two or three of the most drunken. Here Boulton had great difficulty in restraining him. Much had to be endured, and occasional bouts of drunkenness overlooked, although serious accidents resulted. At last two men appeared whose services proved invaluable—Murdoch, already mentioned, and Law—one of whom became famous. Watt was absent when the former called and asked Boulton for employment. The young Scot was the son of a well-known millwright near Ayr who had made several improvements. His famous son worked with him, but being ambitious and hearing of the fame of Boulton and Watt, he determined to seek entrance to Soho works and learn the highest order of handicraft. Boulton had told him that there was at present no place open, but noticing the strange cap the awkward young man had been dangling in his hands, he asked what it was made of. "Timmer," said the lad. "What, out of wood?" "Yes." "*How* was it made?" "I turned it mysel' in a bit lathey o' my own making." This was enough for that rare judge of men. Here was a natural-born mechanic, certain. The young man was promptly engaged for two years at fifteen shillings per week when in shop, seventeen shillings when abroad, and eighteen shillings when in London. His history is the usual march upward until he became his employers' most trusted manager in all their mechanical operations. While engaged upon one critical job, where the engine had defied previous attempts to put it to rights, the people in the house where Murdoch lodged were awakened one night by heavy tramping in his room over-head. Upon entering, Murdoch was seen in his bed clothes heaving away at the bed post in his sleep, calling out "Now she goes, lads, now she goes." His heart was in his work. He had a mission, and only one—to make that engine go.

Of course he rose. There's no holding down such a "dreamer" anywhere in this world. It was not only that he had zeal, for he had sense with it, and was not less successful in conquering the rude Cornishmen who had baffled, annoyed and intimidated Watt. He won their hearts. His ability did not end with curing the defects of machinery; he knew how to manage men. At first he had to depend upon his physical powers. He was an athlete not indisposed to lead the strenuous life. He had not been very long in Cornwall before half a dozen of the mining captains, a class that had tormented poor, retiring and modest Watt, entered the engine-room and began their bullying tricks on him. The Scotch blood was up, Murdoch quietly locked the door and said to the captains, "Now then gentlemen, you shall not leave until we have settled matters once for all." He selected the biggest Cornishman and squared off. The contest was soon over. Murdoch vanquished the bully and was ready for the next. The captains, seeing the kind of man he was, offered terms of peace, hands were shaken all round and they parted good friends, and remained so. We are past that rude age. The skilled, educated manager of to-day can use no weapon so effectively with skilled men as the supreme force of gentleness, the manner, language and action of the educated man, even to the calm, low voice never raised to passionate pitch. He conquers and commands others because he has command of himself.

We must not lose sight of Murdoch. In addition to his rare qualities, he possessed mechanical genius. He was the inventor of lighting by gas, and it was he who made the first model of a locomotive. There was no emergency with engines, no accident, no blunder, but Murdoch was called in. We read with surprise that his wages even in 1780 were only five dollars per week. He then modestly asked for an advance, but this was not given. A present of one hundred dollars, however, was made to him in recognition of his unusual services. Probably the explanation of the failure to increase his wages at the time was that, owing to the condition of the business, no rise in wages could be made to one which would involve an advance to others. Murdoch remained loyal to the firm, however, although invited into partnership by another. Afterward he received due reward. He had always a strong aversion to partnership, no doubt well founded in this case, for during many years failure seemed almost as likely as success. Watt has much to say in his letters about "William" (Murdoch), who, more than anyone, relieved him from trouble.]

The bargainings with mine-owners brought on intense heartaches and broke Watt down completely. Boulton had to go to him again in Cornwall in the autumn of 1779, and as usual succeeded in adjusting many disputes by wise compromises with the grasping owners which Watt's strict sense of justice had denied. Many of these had paid no royalties for years, others disputed Watt's unerring register of fuel consumption (another of his most ingenious inventions now in general use for many purposes), a more heinous offense in his eyes than that of non-payment. "The rascality of man," he writes, "is almost beyond belief." He never was more despondent or more irritable than now. No one was better aware of his weakness than himself. In short, his heartaches and nervousness unfitted him for business. As usual, he attributed his discouragement chiefly to his financial obligations. The firm was as hard pressed as ever. Indeed a new source of danger had developed. Fothergill's affairs became involved, and had it not been for Boulton's capital and credit, the firm of Boulton and Fothergill could not have maintained payment. This had caused a drain upon their resources. Boulton sold the estate which had come to him by his wife, and the greater part of his father's property, and mortgaged the remainder. It is evident that the great captain had taken in hand far too many enterprises. Probably he had not heard the new doctrine: "Put all your eggs in one basket and then watch that basket." He had even ventured considerable sums in blockade running during the American Revolutionary War. It was not without good reason, therefore, that the more cautious Scot addressed to him so many pathetic letters: "I beg of you to attend to these money matters. I cannot rest in my bed until they have some determinate form." Watt's inexperience in money matters caused apprehensions of ruin to arise whenever financial measures were discussed. He was at this time utterly wretched, and Mrs. Watt at last became anxious, long and bravely as she had hitherto borne up and striven to dispel her husband's fears. Never before had she ventured to speak to Boulton upon the subject. She now broke the silence and wrote him in Cornwall a touching letter, stating that her husband's health and spirits had become much worse since Boulton had left Soho. "I know there are several things that so prey upon his mind as to render him perfectly miserable. They never cross his mind, but he is rendered unfit to do anything for a long time." She describes these financial demons that torment him and begs that her writing should not be told to Watt, as it might only add to his troubles. The appeal brings Mrs. Watt before us in a most engaging light.

A study of the problem was made upon Boulton's return and he agreed to close two departments of the business which were so far unprofitable, thus entering upon the right path. The engine having proved itself indispensable, the demand for it was becoming great and pressing from various countries. To concentrate upon its manufacture was obviously the true policy. The great captain's enterprise was not often expended upon failures, and it is with pleasure we find that among the profitable branches which Boulton had encouraged Watt in introducing at Soho, was the copying-press, which Watt invented in 1778, and which we use to this day. In July of that year he writes Dr. Black that he has "lately discovered a method of copying writing instantaneously, provided it has been written within twenty-four hours. I send you a specimen and will impart the secret if it will be of any use to you. It enables me to copy all my business letters." He kept this secret for two years, and in May, 1780, secured a patent after he had completed details of the press and experimented with the ink. One hundred and fifty were made and sold. Thirty of these went abroad. It steadily made its way. Watt, writing some thirty years later, said it had proved so useful to him that it was well worth all the trouble of perfecting it, even if it brought no profit.

We think of Watt and the steam engine appears. Let us however note the unobtrusive little copying-press on the table at his side. Extremes meet here. It would be difficult to name an invention more universally used, in all offices where man labors in any field of activity. In the list of modest inventions of greatest usefulness, the modern copying-press must take high rank, and this we owe entirely to Watt.

Of the same period as the copying-machine is his invention of a drying-machine for cloth, consisting of three cylinders of copper over which the cloth must turn over and under while cylinders are filled with steam, the cloth to be alternately wound off and on the two wooden rollers, by which means it will pass over three cylinders in succession. This machine was erected for Watt's father-in-law, Mr. MacGregor in Glasgow, by an ingenious mechanic, John Gardiner, often employed by Watt in earlier years. "This I apprehend," he writes to David Brewster in 1814, "to be the original from which such machines were made." When we consider the extent to which such steam drying-machines are used in our day, our estimate of the credit due to Watt cannot be small. The drying-machine is no unfit companion to the copying-machine.

Watt revisited Cornwall in 1781 to make an inspection of all the engines. Much he found needing attention and improvement. His evenings were spent designing "road steam-carriages." This was before the day of railroads, and the carriages were to be driven by steam over the ordinary coach roads. He filled a quarto drawing-book with different plans for these, and covered the idea in one of his patent specifications. Boulton suggested in 1781 that the idea of rotary motion should be developed, which Watt had from the first regarded as of prime importance. It was for this he had invented his original wheel engine, and in his first patent of 1769 he describes one method of securing it. It occurred to him that the ordinary engine might be adapted to give the rotary motion. He wrote from Cornwall to Boulton: "As to the circular motion, I will apply it as soon as I can." He prepared a model upon his return to Soho, using a crank connected with the working-beam of the engine for that purpose, which worked satisfactorily. There was nothing new in the crank motion; it was used on every spinning-wheel, grind-stone and foot-lathe turned by hand, but its application to the steam-engine was new. As early as 1771, he writes:

I have at times had my thoughts a good deal upon the subject. In general, it appears to me that a crank of a sufficient sweep will be by much the sweetest motion, and perhaps not the dearest, if its durability be considered ... I then resolved to adopt the crank ... Of this I caused a model to be made, which performed to satisfaction. But being then very much engaged with other business, I neglected to take a patent immediately, and having employed a blackguard of the name of Cartwright (who was afterward hanged), about this model, he, when in company with some of the same sort who worked at Wasborough's mill, and were complaining of its irregularities and frequent disasters, told them he could put them in a way to make a rotative motion which would not go out of order nor stun them with its noise, and accordingly explained to them what he had seen me do. Soon after which, John Steed, who was engineer at Wasborough's mill, took a patent for a rotative motion with a crank, and applied it to their engine. Suspicions arising of Cartwright's treachery, he was strictly questioned, and confessed his part in the transaction when too late to be of service to us.

Overtures were made by Wasborough to exchange patents and work together, which Watt scornfully rejected. He writes:

Though I am not so saucy as many of my countrymen, I have enough innate pride to prevent me from doing a mean action because a servile prudence may dictate it ... I will never meanly sue a thief to give me my own again unless I have nothing left behind.

His blood was up. No dealings with rascals!

July, 1781, Watt had finished his studies, went to Penryn, and swore he had "invented certain new methods of applying the vibrating or reciprocating motion of steam or fire engines to produce a continued rotation or circular motion round an axis or centre, and thereby to give motion to the wheels of mills or other machines."

Watt proceeded to work out the plan of the rotary engine, stimulated by numerous inquiries for steam engines for driving all kinds of mills. He found that "the people in London, Manchester and Birmingham are steam-mill mad."

During many long years of trial with their financial troubles, inferior and drunken workmen, disappointing engines, Cornish mine-owners to annoy him, it is highly probable that Watt only found relief in retiring to his garret to gratify his passion for solving difficult mechanical problems. We may even imagine that from his serious mission—the development of the engine—which was ever present, he sometimes flew to the numerous less exhausting inventions for recreation, as the weary student flies to fiction. His mind at this period seems never to have been at rest. His voluminous correspondence constantly reveals one invention after another upon which he was engaged. A new micrometer, a dividing screw, a new surveying-quadrant, problems for clearing the observed distance of the moon from a star of the effects of refraction and parallax, a drawing-machine, a copying-machine for sculpture—anything and everything he used or saw seems immediately to have been subjected to the question: "Cannot this be improved?" usually with a response in the affirmative.

As we have read, he had long studied the question of a locomotive steam carriage. In Muirhead's Biography, several pages are devoted to this. In his seventh "new improvement," in his patent of 1784, he describes "the principle and construction of steam engines which are applied to give motion to wheel carriages for removing persons, goods, or other matter from place to place, in which case the engines themselves must be portable." Mr. Murdoch made a model of the engine here specified which performed well, but nothing important came of all this until 1802, when the problem was instantly changed by Watt's friend, Mr. Edgeworth, writing him, "I have always thought that steam would become the universal lord, and that we should in time scorn post-horses. *An iron railroad would be a cheaper thing than a road of the common construction.*" Here lay in a few words the idea from which our railway system has sprung. Surely Edgeworth deserves to be placed among the immortals.] As in the case of the steamship, however, the indispensable steam engine of Watt had to furnish the motive power. The railroad is only the necessary smooth track upon which the steam engine could perform its miracle. It is significant that steam power upon roads required the abandonment of the usual highway. So we may believe is the automobile to force new roads of its own, or to widen existing highways, rendering those safe under certain rules for speed of twenty miles per hour, or even more, when they were intended only for eight or ten.

The reading lamp of Watt's day was a poor affair, and as he never saw an inefficient instrument without studying its improvement, he produced a new lamp. He wrote Argand of the Argand burner upon the subject and for a long time Watt lamps were made at the Soho works, which gave a light surpassing in steadiness and brilliance anything of the kind that had yet appeared. He gives four plans for lamps, "with the reservoir below and the stem as tall as you please." He also made an instrument for determining the specific gravity of liquids, and a year after this he "found out a method of working tubes of the elastic resin without dissolving it." The importance of such tubes for a thousand purposes in the arts and sciences is now appreciated.

Watt gave much time to an arithmetical machine which he found exceedingly simple to plan, but he adds, "I have learnt by experience that in mechanics many things fall out between the cup and the mouth." He describes what it is to accomplish, but it remained for Babbage at a much later date to perfect the machine. A machine for copying sculpture amused him for a time but it was never finished.

If any difficulty of a mechanical nature arose, people naturally turned to Watt for a solution. Thus the Glasgow University failed to get pipes for conveying water across the Clyde to stand, the channel of the river being covered with mud and shifty sand, full of inequalities, and subject to the pressure of a considerable body of water. Application was at last made to the recognised genius. If he could not solve it, who could? This was just one of the things that Watt liked to do. He promptly devised an articulated suction pipe with parts formed on the principle of a lobster's tail. This crustacean tube a thousand feet long solved the matter. Watt stated that his services were induced solely by a desire to be of use in procuring good water to the city of Glasgow, and to promote the prosperity of a company which had risked so much for the public good. These were handsomely acknowledged by the presentation to him of a valuable piece of plate.

As another proof of Watt's habit of thinking of everything that could possibly be improved, it may be news to many readers that the consumption of the smoke from steam engines early attracted his attention, and that he patented devices for this. These have been substantially followed in the numerous attempts which have been made from time to time to reduce the huge volumes of smoke that keep so many cities under a cloud. He was successful and his son James writes to him in 1790 from Manchester:

It is astonishing what an impression the smoke-consuming power of the engine has made upon everybody hereabouts. They scarcely trusted to the evidence of their senses. You would be diverted to hear the strange hypotheses which have been stated to account for it.

This is all very well. It is certain that most of the smoke made in manufacturing concerns can be consumed, if manufacturers are compelled by law to erect sufficient heating surface and to include the well-known appliances, including those for careful firing, but no city so far as the writer knows has ever been able to enforce effective laws. There remain the dwellings of the people to deal with, which give forth smoke in large cities in the aggregate far exceeding that made by the manufacturing plants. New York pursues the only plan for ensuring the clearest skies of any large city in the world where coal is generally used, by making the use of bituminous coal unlawful. The enormous growth of present New York (45 per cent. in last decade) is not a little dependent upon theattraction of clear blue sides and the resulting cleanliness of all things in and about the city compared with others. When, by the progress of invention or new methods of distributing heat, smoke is banished, as it probably will be some day, many rich citizens will remain in their respective western cities instead of flocking to the clear blue-skied metropolis, as they are now so generally doing.

Such were some of Watt's by-products. His recreation, if found at all, was found in change of occupation. We read of no idle days, no pleasure trips, no vacations, only change of work.

Rumors of new inventions of engines far excelling his continued to disturb Watt, and much of his time was given to investigation. He thought of a caloric air engine as possibly one of the new ideas; then of the practicability of producing mechanical power by the absorption and condensation of gas on the one hand and by its disengagement and expansion on the other. His mind seemed to range over the entire field of possibilities.

The Hornblower engine had been heralded as sure to displace the Watt. When it was described, it proved to be as Watt said, "no less than our double-cylinder engine, worked upon our principle of expansion. It is fourteen years since I mentioned it to Mr. Smeaton." Watt had explained to Dr. Small his method of working steam expansively as early as May, 1769, and had adopted it in the Soho engine and also in the Shadwell engine erected in that year.

We have seen before that Watt had to retrace his steps and abandon for a time in later engines what he had before ventured upon.

The application of steam for propelling boats upon the water was, at this time (1788), attracting much attention. Boulton and Watt were urged to undertake experiments. This they declined to entertain, having their facilities fully employed in their own field, but finally Fulton, on August 6, 1803, ordered an engine from them from his own drawings, intended for this purpose, repeating the order in person in 1804. It was shipped to America early in 1805, and in 1807 placed upon the Clermont, which ran upon the Hudson River as a passenger boat, attaining a speed of about five miles an hour. This was the first steamboat that was ever used for passengers, and altho Fulton neither invented the boat nor the engine, nor the combination of the two, still he is entitled to great credit for overcoming innumerable difficulties sufficient to discourage most men. Fulton, who was the son of a Scotsman from Dumfrieshire, visited Syminton's steamboat, the *Charlotte Dundas*, in Scotland, in 1801, and had seen it successfully towing canal boats upon the Forth and Clyde Canal. This was the first boat ever propelled by steam successfully for commercial purposes. It was subsequently discarded, not because it did not tow the canal boats, but because the revolving paddle-wheels caused waves that threatened to wash away the canal banks.

Several engines were sent to New York. The men in charge of one found on shipboard a pattern-maker going to America named John Hewitt. He settled in America January 12th, 1796, and became the father of the late famous and deeply lamented Hon. Abram S. Hewitt, long a member of Congress and afterward mayor of New York, foremost in many improvements in the city, the last being the Subway, just opened, which owes its inception to him. For this service, the Chamber of Commerce presented him with a memorial medal. Mr.

Hewitt married a daughter of Peter Cooper, founder of the Cooper Institute, which owes its wonderful development chiefly to him. His children devote themselves and their fortunes to its management. At the time of his death in 1902, he was pronounced "the first private citizen of the Republic." Small engine-shops (of which the ruins still remain), called "Soho" after their prototype, were erected by his father near New York City, on the Greenwood division of the Erie Railroad. The railroad station was called "Soho" by Mr. Abram S. Hewitt, who was then president of the railroad company. Upon Mr. Hewitt's eightieth birthday congratulations poured in from all quarters. One cable from abroad attracted attention as appropriate and deserved: "Ten octaves every note truly struck and grandly sung." No man in private life passed away in our day with such general lamentation. The Republic got even more valuable material than engines from the old home in the ship that arrived on January 12, 1796.

We must not permit ourselves to forget that it was not until the Watt engine was applied to steam navigation that the success of the latter became possible. It was only by this that it could be made practicable, so that the steamship is the product of the steam-engine, and it is to Watt we owe the modern twenty-three-thousand-ton monster (and larger monsters soon to come), which keeps its course against wind and tide, almost "unshaked of motion," for this can now properly be said. Passengers crossing the Atlantic from port to port now scarcely know anything of irregular motion, and never more than the gentlest of slight heaves, even during the gale that

"Catches the ruffian billows by their tops,Curling their monstrous heads."

The ocean, traversed in these ships, is a smooth highway—nothing but a ferry—and a week spent upon it has become perhaps the most enjoyable and the most healthful of holiday excursions, provided the prudent excursionist has left behind positive instructions that wireless telegrams shall not follow.

CHAPTER VII

SECOND PATENT

The number and activity of rivals attracted to the steam engine and its possible improvement, some of whom had begun infringements upon the Watt patents, alarmed Messrs. Watt and Boulton so much that they decided Watt should apply for another patent, covering his important improvements since the first. Accordingly, October 25, 1781, the patent (already referred to on p. 91) was secured, "for certain new methods of producing a continued rotative motion around an axis or centre, and thereby to give motion to the wheels of mills or other machines."

This patent was necessary in consequence of the difficulties experienced in working the steam wheels or rotatory engines described in the first patent of 1769, and by Watt's having been so unfairly anticipated, by Wasborough in the crank motion.

No less than five different methods for rotatory motion are described in the patent, the fifth commonly known as the "sun and planet wheels," of which Watt writes to Boulton, January 3, 1782,

I have tried a model of one of my old plans of rotative engines, revived and executed by Mr. Murdoch, which merits being included in the specification as a fifth method; for which purpose I shall send a drawing and description next post. It has the singular property of going twice round for each stroke of the engine, and may be made to go oftener round, if required, without additional machinery.

Then followed an explanation of the sketch which he sent, and two days later he wrote, "I send you the drawings of the fifth method, and thought to have sent you the description complete, but it was late last night before I finished so far, and to-day have a headache, therefore only send you a rough draft of part."

In all of these Watt recommended that a fly-wheel be used to regulate the motion, but in the specification for the patent of the following year, 1782, his double-acting engine produced a more regular motion and rendered a fly-wheel unnecessary, "so that," he says, "in most of our great manufactories these engines now supply the place of water, wind and horse mills, and instead of carrying the work to the power, the prime agent is placed wherever it is most convenient to the manufacturer."

This marks one of the most important stages in the development of the steam engine. It was at last the portable machine it remains to-day, and was placed wherever convenient, complete in itself and with the rotative motion adaptable for all manner of work. The ingenious substitutes Watt had to invent to avoid the obviously perfect crank motion have of course all been discarded, and nothing of these remains except as proofs, where none are needed, that genius has powers in reserve for emergencies; balked in one direction, it hews out another path for itself.

While preparing the specification for this patent of 1781, Watt was busy upon another specification quite as important, which appeared in the following year, 1782. It embraced the following new improvements, the winnowing of numberless ideas and experiments that he had conceived and tested for some years previous:

1. The use of steam on the expansive principle; together with various methods or contrivances (six in number, some of them comprising various modifications), for equalising the expansive power.

2. The double-acting engine; in which steam is admitted to press the piston upward as well as downward; the piston being also aided in its ascent as well as in its descent by a vacuum produced by condensation on the other side.

3. The double-engine; consisting of two engines, primary and secondary, of which the steam-vessels and condensers communicate by pipes and valves, so that they can be worked either independently or in concert; and make their strokes either alternately or both together, as may be required.

4. The employment of a toothed rack and sector, instead of chains, for guiding the piston-rod.

5. A rotative engine, or steam-wheel.

Here we have three of the vital elements required toward the completion of the work: first, steam used expansively; second, the double-acting engine. It will be remembered that Watt's first engines only took in steam at the bottom of the cylinder, as Newcomen's did, but with this difference: Watt used the steam to perform work which Newcomen could not do, the latter only using steam to force the piston itself upward. Now came Watt's great step forward. Having a cylinder closed at the top, while the Newcomen cylinder remained open, it was as easy to admit steam at the top to press the piston down as to admit it at the bottom to press the piston up; also as easy to apply his condenser to the steam above as below, at the moment a vacuum was needed. All this was ingeniously provided for by numerous devices and covered by the patent. Third, he went one step farther to the compound engine, consisting of two engines, primary and secondary, working steam expansively independently or in concert, with strokes alternate or simultaneous. The compound engine was first thought of by Watt about 1767. He laid a large drawing of it on parchment before parliament when soliciting an extension of his first patent. The reason he did not proceed to construct it was "the difficulty he had encountered in teaching others the construction and use of the single engine, and in overcoming prejudices"; the patent of 1782 was only taken out because he found himself "beset with a host of plagiaries and pirates."

One of the earliest of these double-acting engines was erected at the Albion Mills, London, in 1786. Watt writes:

The mention of Albion Mills induces me to say a few words respecting an establishment so unjustly calumniated in its day, and the premature destruction of which, by fire, in 1791, was, not improbably, imputed to design. So far from being, as misrepresented, a monopoly injurious to the public, it was the means of considerably reducing the price of flour while it continued at work.

The "double-acting" engine was followed by the "compound" engine, of which Watt says:

A new compound engine, or method of connecting together the cylinders and condensers of two or more distinct engines, so as to make the steam which has been employed to press on the piston of the first, act expansively upon the piston of the second, etc., and thus derive an additional power to act either alternately or co-jointly with that of the first cylinder.

We have here, in all substantial respects, the modern engine of to-day.

Two fine improvements have been made since Watt's time: first, the piston-rings of Cartwright, which effectively removed one of Watt's most serious difficulties, the escape of steam, even though the best packing he could devise were used—the chief reason he could not use high-pressure steam. In our day, the use of this is rapidly extending, as is that of superheated steam. Packing the piston was an elaborate operation even after Watt's day.

It was not because Watt did not know as well as any of our present experts the advantages of high pressures, that he did not use them, but simply because of the mechanical difficulties then attending their adoption. He was always in advance of mechanical practicalities rather than behind, and as we have seen, had to retrace his steps, in the case of expansion.

The other improvement is the cross-head of Haswell, an American, a decided advance, giving the piston rod a smooth and straight bed to rest upon and freeing it from all disturbance. The drop valve is now displacing the slide valve as a better form of excluding or admitting steam.

Watt of course knew nothing of the thermo-dynamic value of high temperature without high pressure, altho fully conversant with the value of pressures. This had not been even imagined by either philosopher or engineer until discovered by Carnot as late as 1824. Even if he had known about it the mechanical arts in his day were in no condition to permit its use. Even high pressures were impracticable to any great extent. It is only during the past few years that turbines and superheating, having long been practically discarded, show encouraging signs of revival. They give great promise of advancement, the hitherto insuperable difficulties of lubrication and packing having been overcome within the last five years. Superheating especially promises to yield substantial results as compared with the practice with ordinary engines, but the margin of saving in steam over the best quadruple expansion engine cannot be great. Lord Kelvin however expects it to be the final contribution of science to the highest possible economy in the steam engine.

In the January (1905) number of "Stevens Institute Indicator," Professor Denton has an instructive résumé of recent steam engine economics. He tells us that Steam Turbines are now being applied to Piston Engines to operate with the latter's exhaust, to effect the same saving as the sulphur dioxide cylinder; and adds

that the Turbine is a formidable competitor to the Piston Engine is mainly due to the fact that it more completely realizes the expansive principle enunciated in the infancy of steam history as the fundamental factor of economy by its sagacious founder, the immortal Watt.

Watt's favorite employment in Soho works late in 1783 and early in 1784 was to teach his engine, now become as docile as it was powerful, to work a tilt hammer. In 1777 he had written Boulton that

Wilkinson wants an engine to raise a stamp of 15 cwt. thirty or forty times in a minute. I have set Webb to work to try it with the little engine and a stamp-hammer of 60 lbs. weight. Many of these *battering rams* will be wanted if they answer.

The trial was successful. A new machine to work a 700 lbs. hammer for Wilkinson was made, and April 27, 1783, Watt writes that it makes from 15 to 50, and even 60, strokes per minute, and works a hammer, raised two feet high, which has struck 300 blows per minute.

The engine was to work two hammers, but was capable of working four of 7 cwt. each. He says, with excusable pride,

I believe it is a thing never done before, to make a hammer of that weight make 300 blows per minute; and, in fact, it is more amatter to brag of than for any other use, as the rate wanted is from 90 to 100 blows, being as quick as the workmen can manage the iron under it.

This most ingenious application of steam power was included in Watt's next patent of April 28, 1784. It embraced many improvements, mostly, however, now of little consequence, the most celebrated being "parallel motion," of which Watt was prouder than any other of his triumphs. He writes to his son, November, 1808, twenty-four years after it was invented (1784):

Though I am not over anxious after fame, yet I am more proud of the parallel motion than of any other mechanical invention I have ever made.

He wrote Boulton, in June, 1784:

I have started a new hare. I have got a glimpse of a method of causing a piston-rod to move up and down perpendicularly, by only fixing it to a piece of iron upon the beam ... I think it one of the most ingenious simple pieces of mechanism I have contrived.

October, 1784, he writes:

The new central perpendicular motion answers beyond expectation, and does not make the shadow of a noise.

He says:

When I saw it in movement, it afforded me all the pleasure of a novelty, as if I had been examining the invention of another.

When beam-engines were universally used for pumping, this parallel motion was of great advantage. It has been superseded in our day, by improved piston guides and cross-heads, the construction of which in Watt's day was impossible, but no invention has commanded in greater degree the admiration of all who comprehend the principles upon which it acts, or who have witnessed the smoothness, orderly power and "sweet simplicity" of its movements. Watt's pride in it as his favorite invention in these respects is fully justified.

A detailed specification for a road steam-carriage concludes the claims of this patent, but the idea of railroads, instead of common roads, coming later left the construction of the locomotive to Stephenson.]

Watt's last patent bears date June 14, 1785, and was

for certain newly improved methods of constructing furnaces or fire-places for heating, boiling, or evaporating of water and other liquids which are applicable to steam engines and other purposes, and also for heating, melting, and smelting of metals and their ores, whereby greater effects are produced from the fuel, and the smoke is in a great measure prevented or consumed.

The principle, "an old one of my own," as Watt says, is in great part acted upon to-day.

So numerous were the improvements made by Watt at various periods, which greatly increased the utility of his engine, it would be in vain to attempt a detailed recital of his endless contrivances, but we may mention as highly important, the throttle-valve, the governor, the steam-gauge and the indicator. Muirhead says:

The throttle-valve is worked directly by the engineer to start or stop the engine, and also to regulate the supply of steam. Watt describes it as a circular plate of metal, having a spindle fixed across its diameter, the plate being accurately fitted to an aperture in a metal ring of some thickness, through the edgeway of which the spindle is fitted steam-tight, and the ring fixed between the two flanches of the joint of the steam-pipe which is next to the cylinder. One end of the spindle, which has a square upon it, comes through the ring, and has a spanner fixed upon it, by which it can be turned in either direction. When the valve is parallel to the outsides of the ring, it shuts the opening nearly perfectly; but when its plane lies at an angle to the ring, it admits more or less steam according to the degree it has opened; consequently the piston is acted upon with more or less force.

Papin preferred gunpowder as a safer source of power than steam, but that was before it had been automatically regulated by the "Governor." The governor has always been the writer's favorite invention, probably because it was the first he fully understood. It is an application of the centrifugal principle adapted and mechanically improved. Two heavy revolving balls swing round an upright rod. The faster the rod revolves the farther from it the balls swing out. The slower it turns the closer the balls fall toward it. By proper attachments the valve openings admitting steam are widened or narrowed accordingly. Thus the higher speed of the engine, the less steam admitted, the slower the speed the more steam admitted. Hence any uniform speed desired can be maintained: should the engine be called upon to perform greater service at one moment than another, as in the case of steel rolling mills, speed being checked when the piece of steel enters the rolls, immediately the valves widen, more steam rushes into the engine, and *vice versa*. Until the governor came regular motion was impossible—steam was an unruly steed.

Arago describes the steam-gauge thus:

It is a short glass tube with its lower end immersed in a cistern of mercury, which is placed within an iron box screwed to the boiler steam-pipe, or to some other part communicating freely with the steam, which, pressing on the surface of the mercury in the cistern, raises the mercury in the tube (which is open to the air at the upper end), and its altitude serves to show the elastic power of the steam over that of the atmosphere.

The indicator he thus describes:

The barometer being adapted only to ascertain the degree of exhaustion in the condenser where its variations were small, the vibrations of the mercury rendered it very difficult, if not impracticable, to ascertain the state of the exhaustion of the cylinder at the different periods of the stroke of the engine; it became therefore necessary to contrive an instrument for that purpose that should be less subject to vibration, and should show nearly the degree of exhaustion in the cylinder at all periods. The following instrument, called the Indicator, is found to answer the end sufficiently. A cylinder about an inch diameter, and six inches long, exceedingly truly bored, has a solid piston accurately fitted to it, so as to slide easy by the help of some oil; the stem of the piston is guided in the direction of the axis of the cylinder, so that it may not be subject to jam, or cause friction in any part of its motion. The bottom of this cylinder has a cock and small pipe joined to it which, having a conical end, may be inserted in a hole drilled in the cylinder of the engine near one of the ends, so that, by opening the small cock, a communication may be effected between the inside of the cylinder and the indicator.

The cylinder of the indicator is fastened upon a wooden or metal frame, more than twice its own length; one end of a spiral steel spring, like that of a spring steel-yard, is attached to the upper part of the frame, and the other end of the spring is attached to the upper end of the piston-rod of the indicator. The spring is made of such a strength, that when the cylinder of the indicator is perfectly exhausted, the pressure of the atmosphere may force its piston down within an inch of its bottom. An index being fixed to the top of its piston-rod, the point where it stands, when quite exhausted, is marked from an observation of a barometer communicating with the same exhausted vessel, and the scale divided accordingly.

Improvements come in many ways, sometimes after much thought and after many experimental failures. Sometimes they flash upon clever inventors, but let us remember this is only after they have spent long years studying the problem. In the case of the steam engine, however, a quite important improvement came very curiously. Humphrey Potter was a lad employed to turn off and on the stop cocks of a Newcomen engine, a monotonous task, for, at every stroke one had to be turned to let steam into the boiler and another for injecting the cold water to condense it, and this had to be done at the right instant or the engine could not move. How to relieve himself from the drudgery became the question. He wished time to play with the other boys whose merriment was often heard at no great distance, and this set him thinking. Humphrey saw that the beam in its movements might serve to open and shut these stop cocks and he promptly began to attach cords to the cocks and then tied them at the proper points to the beam, so that ascending it pulled one cord and descending the other. Thus came to us perhaps not the first automatic device, but no doubt the first of its kind that was ever seen there. The steam engine henceforth was self-attending, providing itself for its own supply of steam and for its condensation with perfect regularity. It had become in this feature automatic.

The cords of Potter gave place to vertical rods with small pegs which pressed upward or downward as desired. These have long since been replaced by other devices, but all are only simple modifications of a contrivance devised by the mere lad whose duty it was to turn the stop cocks.

It would be interesting to know the kind of man this precocious boy inventor became, or whether he received suitable reward for his important improvement. We search in vain; no mention of him is to be found. Let us, however, do our best to repair the neglect and record that, in the history of the steam engine, Humphrey Potter must ever be honorably associated with famous men as the only famous boy inventor.

In the development of the steam engine, we have one purely accidental discovery. In the early Newcomen engines, the head of the piston was covered by a sheet of water to fill the spaces between the circular contour of the movable piston and the internal surface of the cylinder, for there were no cylinder-boring tools in those days, and surfaces of cylinders were most irregular. To the surprise of the engineer, the engine began one day working at greatly increased speed, when it was found that the piston-head had been pierced by accident and that the cold water had passed in small drops into the cylinder and had condensed the steam, thus rapidly making a more perfect vacuum. From this accidental discovery came the improved plan of injecting a shower of cold water through the cylinder, the strokes of the engine being thus greatly increased.

The year 1783 was one of Watt's most fruitful years of the dozen which may be said to have teemed with his inventions. His celebrated discovery of the composition of water was published in this year. The attempts made to deprive him of the honor of making this discovery ended in complete failure. Sir Humphrey Davy, Henry, Arago, Liebig, and many others of the highest authority acknowledged and established Watt's claims.

The true greatness of the modest Watt was never more finely revealed than in his correspondence and papers published during the controversy. Watt wrote Dr. Black, April 21st, that he had handed his paper to Dr. Priestley to be read at the Royal Society. It contained the new idea of water, hitherto considered an element and now discovered to be a compound. Thus was announced one of the most wonderful discoveries found in the history of science. It was justly termed the beginning of a new era, the dawn of a new day in physical chemistry, indeed the real foundation for the new system of chemistry, and, according to Dr. Young, "a discovery perhaps of greater importance than any single fact which human ingenuity has ascertained either before or since." What Newton had done for light Watt was held to have done for water? Muirfield well says:

It is interesting in a high degree to remark that for him who had so fully subdued to the use of man the gigantic power of steam it was also reserved to unfold its compound natural and elemental principles, as if on this subject there were to be nothing which his researches did not touch, nothing which they touched that they did not adorn.

Arago says:

In his memoir of the month of April, Priestley added an important circumstance to those resulting from the experiments of his predecessors: he proved that the weight of the water which is deposited upon the sides of the vessel, at the instant of the detonation of the oxygen and hydrogen, is precisely the same as the weights of the two gases.

Watt, to whom Priestley communicated this important result, immediately perceived that proof was here afforded that water was not a simple body. Writing to his illustrious friend, he asks:

What are the products of your experiment? They are *water, light* and *heat*. Are we not, thence, authorised to conclude that water is a compound of the two gases, oxygen and hydrogen, deprived of a portion of their latent or elementary heat; that oxygen is water deprived of its hydrogen, but still united to its latent heat and light? If light be only a modification of heat, or a simple circumstance of its manifestation, or a component part of hydrogen, oxygen gas will be water deprived of its hydrogen, but combined with latent heat.

This passage, so clear, so precise, and logical, is taken from a letter of Watt's, dated April 26, 1783. The letter was communicated by Priestley to several of the scientific men in London, and was transmitted immediately afterward to Sir Joseph Banks, the President of the Royal Society, to be read at one of the meetings of that learned body.

Watt had for many years entertained the opinion that air was a modification of water. He writes Boulton, December 10, 1782:

You may remember that I have often said, that if water could be heated red-hot or something more, it would probably be converted into some kind of air, because steam would in that case have lost all its latent heat, and that it would have been turned solely into sensible heat, and probably a total change of the nature of the fluid would ensue.

A month after he hears of Priestley's experiments, he writes Dr. Black (April 21, 1783) that he "believes he has found out the cause of the conversion of water into air." A few days later, he writes to Dr. Priestley:

In the deflagration of the inflammable and dephlogisticated airs, the airs unite with violence—become red-hot—and, on cooling, totally disappear. The only fixed matter which remains is *water*, and *water*, *light*, and *heat*, are all the products. Are we not then authorised to conclude that water is composed of dephlogisticated and inflammable air, or phlogiston, deprived of part of their latent heat; and that dephlogisticated, or pure air, is composed of water deprived of its phlogiston, and united to heat and light; and if light be only a modification of heat, or a component part of phlogiston, then pure air consists of water deprived of its phlogiston and of latent heat?

It appears from the letter to Dr. Black of April 21st, that Mr. Watt had, on that day, written his letter to Dr. Priestley, to be read by him to the Royal Society, but on the 26th he informs Mr. DeLuc, that having observed some inaccuracies of style in that letter, he had removed them, and would send the Doctor a corrected copy in a day or two, which he accordingly did on the 28th; the corrected letter (the same that was afterward embodied verbatim in the letter to Mr. DeLuc, printed in the Philosophical Transactions), being dated April 26th. In enclosing it, Mr. Watt adds, "As to myself, the more I consider what I have said, I am the more satisfied with it, as I find none of the facts repugnant."

Thus was announced for the first time one of the most wonderful discoveries recorded in the history of science, startling in its novelty and yet so simple.

Watt had divined the import of Priestley's experiment, for he had mastered all knowledge bearing upon the question, but even when this was communicated to Priestley, he could not accept it, and, after making new experiments, he writes Watt, April 29, 1783, "Behold with surprise and indignation the figure of an apparatus that has utterly ruined your beautiful hypothesis," giving a rough sketch with his pen of the apparatus employed. Mark the promptitude of the master who had deciphered the message which the experimenter himself could not translate. He immediately writes in reply May 2, 1783:

I deny that your experiment ruins my hypothesis. It is not founded on so brittle a basis as an earthen retort, nor on *its* converting water into air. I founded it on the other facts, and was obliged to stretch it a good deal before it would fit this experiment.... I maintain my hypothesis until it shall be shown that the water formed after the explosion of the pure and inflammable airs, has some other origin.

He also writes to Mr. DeLuc on May 18th:

I do not see Dr. Priestley's experiment in the same light that he does. It does not disprove my theory.... My assertion was simply, that air (*i.e.*, dephlogisticated air, or oxygen, which was also commonly called vital air, pure air, or simple *air*) was water deprived of its phlogiston, and united to heat, which I grounded on the decomposition of air by inflammation with inflammable air, the residuum, or product of which, is only water and heat.

Having, by experiments of his own, fully satisfied himself of the correctness of his theory, in November he prepared a full statement for the Royal Society, having asked the society to withhold his first paper until he could prove it for himself by experiment. He never doubted its correctness, but some members of the society advised that it had better be supported by facts.

When the discovery was so daring that Priestley, who made the experiments, could not believe it and had to be convinced by Watt of its correctness, there seems little room left for other claimants, nor for doubt as to whom is due the credit of the revelation.

Watt encountered the difficulties of different weights and measures in his studies of foreign writers upon chemistry, a serious inconvenience which still remains with us.

He wrote Mr. Kirwan, November, 1783:

I had a great deal of trouble in reducing the weights and measures to speak the same language; and many of the German experiments become still more difficult from their using different weights and different divisions of them in different parts of that empire. It is therefore a very desirable thing to have these difficulties removed, and to get all philosophers to use pounds divided in the same manner, and I flatter myself that may be accomplished if you, Dr. Priestley, and a few of the French experimenters will agree to it; for the utility is so evident, that every thinking person must immediately be convinced of it.

Here follows his plan: Let the

Philosophical pound consist of 10 ounces, or 10,000 grains.
The ounce   "   " 10 drachms or 1,000   "
the drachm "   " 100 grains.

Let all elastic fluids be measured by the ounce measure of water, by which the valuation of different cubic inches will be avoided, and the common decimal tables of specific gravities will immediately give the weights of those elastic fluids.

If all philosophers cannot agree on one pound or one grain, let every one take his own pound or his own grain; it will affect nothing but doses of medicines, which must be corrected as is now done; but as it would be much better that the identical pound was used by all. I would propose that the Amsterdam or Paris pound be assumed as the standard, being now the most universal in Europe: it is to our avoirdupois pound as 109 is to 100. Our avoirdupois pound contains 7,000 of our grains, and the Paris pound 7,630 of our grains, but it contains 9,376 Paris grains, so that the division into 10,000 would very little affect the Paris grain. I prefer dividing the pound afresh to beginning with the Paris grain, because I believe the pound is very general, but the grain local.

Dr. Priestley has agreed to this proposal, and has referred it to you to fix upon the pound if you otherwise approve of it. I shall be happy to have your opinion of it as soon as convenient, and to concert with you the means of making it universal.... I have some hopes that the foot may be fixed by the pendulum and a measure of water, and a pound derived from that; but in the interim let us at least assume a proper division, which from the nature of it must be intelligible as long as decimal arithmetic is used.

He afterward wrote, in a letter to Magellan:

As to the precise foot or pound, I do not look upon it to be very material, in chemistry at least. Either the common English foot may be adopted according to your proposal, which has the advantage that a cubic foot is exactly 1,000 ounces, consequently the present foot and ounce would be retained; or a pendulum which vibrates 100 times a minute may be adopted for the standard, which would make the foot 14.2 of our present inches, and the cubic foot would be very exactly a bushel, and would weigh 101 of the present pounds, so that the present pound would not be much altered. But I think that by this scheme the foot would be too large, and that the inconvenience of changing all the foot measures and things depending on them, would be much greater than changing all the pounds, bushels, gallons, etc. I therefore give the preference to those plans which retain the foot and ounce.

The war of the standards still rages—metric, or decimal, or no change. What each nation has is good enough for it in the opinion of many of its people. Some day an international commission will doubtless assemble to bring order out of chaos. As far as the English-speaking race is concerned, it seems that a decided improvement could readily be affected with very trifling, indeed scarcely perceptible, changes. Especially is this so with money values. Britain could merge her system with those of Canada and America, by simply making her "pound" the exact value of the American five dollars, it being now only ten pence less; her silver coinage one and two shillings equal to quarter- and half-dollars, the present coin to be recoined upon presentation, but meanwhile to pass current. Weights and measures are more difficult to assimilate. Science being world-wide, and knowing no divisions, should use uniform terms. Alas! at the distance of nearly a century and a half we seem no nearer the prospect of a system of universal weights and measures than in Watt's day, but Watt's idea is not to be lost sight of for all that. He was a seer who often saw what was to come.

We have referred to the absence of holidays in Watt's strenuous life, but Birmingham was remarkable for a number of choice spirits who formed the celebrated Lunar Society, whose members were all devoted to the pursuit of knowledge and mutually agreeable to one another. Besides Watt and Boulton, there were Dr. Priestley, discoverer of oxygen gas, Dr. Darwin, Dr. Withering, Mr. Keir, Mr. Galton, Mr. Wedgwood of Wedgwood ware fame, who had monthly dinners at their respective houses—hence the "Lunar" Society. Dr. Priestley, discoverer of oxygen, who arrived in Birmingham in 1780, has repeatedly mentioned the great pleasure he had in having Watt for a neighbor. He says:

I consider my settlement at Birmingham as the happiest event in my life; being highly favourable to every object I had in view, philosophical or theological. In the former respect I had the convenience of good workmen of every kind, and the society of persons eminent for their knowledge of chemistry; particularly Mr. Watt, Mr. Keir, and Dr. Withering. These, with Mr. Boulton and Dr. Darwin, who soon left us by removing from Lichfield to Derby, Mr. Galton, and afterwards Mr. Johnson of Kenilworth and myself, dined together every month, calling ourselves *the Lunar Society*, because the time of our meeting was near the full-moon—in order,

As he elsewhere says, to have the benefit of its light in returning home.

Richard Lovell Edgeworth says of this distinguished coterie:

By means of Mr. Keir, I became acquainted with Dr. Small of Birmingham, a man esteemed by all who knew him, and by all who were admitted to his friendship beloved with no common enthusiasm. Dr. Small formed a link which combined Mr. Boulton, Mr. Watt, Dr. Darwin, Mr. Wedgwood, Mr. Day, and myself together—men of very different characters, but all devoted to literature and science. This mutual intimacy has never been broken but by death, nor have any of the number failed to distinguish themselves in science or literature. Some may think that I ought with due modesty to except myself. Mr. Keir, with his knowledge of the world and good sense; Dr. Small, with his benevolence and profound sagacity; Wedgwood, with his increasing industry, experimental variety, and calm investigation; Boulton, with his mobility, quick perception, and bold adventure; Watt, with his strong inventive faculty, undeviating steadiness, and bold resources; Darwin, with his imagination, science, and poetical excellence; and Day with his unwearied research after truth, his integrity and eloquence proved altogether such a society as few men have had the good fortune to live with; such an assemblage of friends, as fewer still have had the happiness to possess, and keep through life.

The society continued to exist until the beginning of the century, 1800. Watt was the last surviving member. The last reference is Dr. Priestley's dedication to it, in 1793, of one of his works "Experiments on the Generation of Air from Water," in which he says:

There are few things that I more regret, in consequence of my removal from Birmingham, than the loss of your society. It both encouraged and enlightened me; so that what I did there of a philosophical kind ought in justice to be attributed almost as much to you as to myself. From our cheerful meetings I never absented myself voluntarily, and from my pleasing recollection they will never be absent. Should the cause of our separation make it necessary for to me remove to a still greater distance from you, I shall only think the more, and with the more regret, of our past interviews.... Philosophy engrossed us wholly. Politicians may think there are no objects of any consequence besides those which immediately interest *them*. But objects far superior to any of which they have an idea engaged our attention, and the discussion of them was accompanied with a satisfaction to which they are strangers. Happy would it be for the world if their pursuits were as tranquil, and their projects as innocent, and as friendly to the best interests of mankind, as ours.

That the partners, Boulton and Watt, had such pleasure amid their lives of daily cares, all will be glad to know. It was not all humdrum money-making nor intense inventing. There was the society of gifted minds, the serene atmosphere of friendship in the high realms of mutual regard, best recreation of all.

In 1786, quite a break in their daily routine took place. In that year Messrs. Boulton and Watt visited Paris to meet proposals for their erecting steam engines in France under an exclusive privilege. They were also to suggest improvements on the great hydraulic machine of Marly. Before starting, the sagacious and patriotic Watt wrote to Boulton:

I think if either of us go to France, we should first wait upon Mr. Pitt (prime minister), and let him know our errand thither, that the tongue of slander may be silenced, all undue suspicion removed, and ourselves rendered more valuable in his eyes, because others desire to have us!

They had a flattering reception in Paris from the ministry, who seemed desirous that they should establish engine-works in France. This they absolutely refused to do, as being contrary to the interests of their country. It may be feared we are not quite so scrupulous in our day. On the other hand, refusal now would be fruitless, it has become so easy to obtain plans, and even experts, to build machines for any kind of product in any country. Automatic machinery has almost dispelled the need for so-called skilled labor. East Indians, Mexicans, Japanese, Chinese, all become more or less efficient workers with a few month's experience. Manufacturing is therefore to spread rapidly throughout the world. All nations may be trusted to develop, and if necessary for a time protect, their natural resources as a patriotic duty. Only when prolonged trials have been made can it be determined which nation can best and most cheaply provide the articles for which raw material abounds.

The visit to Paris enabled Watt and Boulton to make the acquaintance of the most eminent men of science, with whom they exchanged ideas afterward in frequent and friendly correspondence. Watt described himself as being, upon one occasion, "drunk from morning to night with Burgundy and undeserved praise." The latter was always a disconcerting draught for our subject; anything but reference to his achievements for the modest self-effacing genius.

While in Paris, Berthollet told Watt of his new method of bleaching by chlorine, and gave him permission to communicate it to his father-in-law, who adopted it in his business, together with several improvements of Watt's invention, the results of a long series of experiments. Watt, writing to Mr. Macgregor, April 27, 1787, says:

In relation to the inventor, he is a man of science, a member of the Academy of Sciences at Paris, and a physician, not very rich, a very modest and worthy man, and an excellent chemist. My sole motives in meddling with it were to procure such reward as I could to a man of merit who had made an extensively useful discovery in the arts, and secondly, I had an immediate view to your interest; as to myself, I had no lucrative views whatsoever, it being a thing out of my way, which both my business and my health prevented me from pursuing further than it might serve for amusement when unfit for more serious business. Lately, by a letter from the inventor, he informs me that he gives up all intentions of pursuing it with lucrative views, as he says he will not compromise his quiet and happiness by engaging in business; in which, perhaps, he is right; but if the discovery has real merit, as I apprehend, he is certainly entitled to a generous reward, which I would wish for the honour of Britain, to procure for him; but I much fear, in the way you state it, that nothing could be got worth his acceptance.

France has been distinguished for men of science who have thus refrained from profiting by their inventions. Pasteur, in our day, perhaps the most famous of all, the liver, not only of the simple but of the ideal life, laboring for the good of humanity—service to man—and taking for himself the simple life, free from luxury, palace, estate, and all the inevitable cares accompanying ostentatious living. Berthollet preceded him. Like Agassiz, these gifted souls were "too busy to make money."

In 1792, when Boulton had passed the allotted three score years and ten, and Watt was over three score, they made a momentous decision which brought upon them several years of deep anxiety. Fortunately the sons of the veterans who had recently been admitted to the business proved of great service in managing the affair, and relieved their parents of much labor and many journeys. Fortunate indeed were Watt and Boulton in their partnership, for they became friends first and partners afterward. They were not less fortunate in each having a talented son, who also became friends and partners like their fathers before them. The decision was that the infringers of their patents were to be proceeded against. They had to appeal to the law to protect their rights.

Watt met the apparently inevitable fate of inventors. Rivals arose in various quarters to dispute his right to rank as the originator of many improvements. No reflection need be made upon most rival claimants to inventions. Some wonderful result is conceived to be within the range of possibility, which, being obtained, will revolutionise existing modes. A score of inventive minds are studying the problem throughout the civilised world. Every day or two some new idea flashes upon one of them and vanishes, or is discarded after trial. One day the announcement comes of triumphant success with the very same idea slightly modified, the modification or addition, slight though this may be, making all the difference between failure and success. The man has arrived with the key that opens the door of the treasure-house. He sets the egg on end perhaps by as obvious a plan as chipping the end. There arises a chorus of strenuous claimants, each of whom had thought of that very device long ago. No doubt they did. They are honest in their protests and quite persuaded in their own minds that they, and not the Watt of the occasion, are entitled to the honor of original discovery. This very morning we read in the press a letter from the son of Morse, vindicating his father's right to rank as the father of the telegraph, a son of Vail, one of his collaborators, having claimed that his father, and not Morse, was the real inventor. The most august of all bodies of men, since its decisions overrule both Congress and President, the Supreme Court of the United States, has shown rare wisdom from its inception, and in no department more clearly than in that regarding the rights of inventors. No court has had such experience with patent claims, for no nation has a tithe of the number to deal with. Throughout its history, the court has attached more and more importance to two points: First, is the invention valuable? Second, who proved this in actual practice? These points largely govern its decisions.

The law expenses of their suits seemed to Boulton and Watt exorbitant, even in that age of low prices compared to our own. One solicitor's bill was for no less than $30,000, which caused Watt years afterward, when speaking of an enormous charge to say that "it would not have disgraced a London solicitor." When we find however, that this was for four years' services, the London solicitor appears in a different light. "In the whole affair," writes Watt to his friend Dr. Black, January 15, 1797, "nothing was so grateful to me as the zeal of our friends and the activity of our young men, which were unremitting."

The first trial ended June 22, 1793, with a verdict for Watt and Boulton by the jury, subject to the opinion of the court as to the validity of the patent. On May 16, 1795, the case came on for judgment, when unfortunately the court was found divided, two for the patent and two against. Another case was tried December 16, 1796, with a special jury, before Lord Chief Justice Eyre; the verdict was again for the plaintiffs. Proceedings on a writ of error had the effect of affirming the result by the unanimous opinion of the four judges, before whom it was ably and fully argued on two occasions.

The testimony of Professor Robison, Watt's intimate friend of youth in Glasgow, was understood to have been deeply impressive, and to have had a decisive effect upon judges and jury.

All the claims of Watt were thus triumphantly sustained. The decision has always been considered of commanding importance to the law of patents in Britain, and was of vast consequence to the firm of Watt and Boulton pecuniarily. Heavy damages and costs were due from the actual defendants, and the large number of other infringers were also liable for damages. As was to have been expected, however, the firm remembered that to be merciful in the hour of victory and not to punish too hard a fallen foe, was a cardinal virtue. The settlements they made were considered most liberal and satisfactory to all. Watt used frequently long afterward to refer to his specifications as his old and well-tried friends. So indeed they proved, and many references to their wonderful efficiency were made.

With the beginning of the new century, 1800, the original partnership of the famous firm of Boulton and Watt expired, after a term of twenty-five years, as did the patents of 1769 and 1775. The term of partnership had been fixed with reference to the duration of the patents. Young men in their prime, Watt at forty and Boulton about fifty when they joined hands, after a quarter-century of unceasing and anxious labor, were disposed to resign the cares and troubles of business to their sons. The partnership therefore was not renewed by them, but their respective shares in the firm were agreed upon as the basis of a new partnership between their sons, James Watt, Jr., Matthew Robinson Boulton and Gregory Watt, all distinguished for abilities of no mean order, and in a great degree already conversant with the business, which their wise fathers had seen fit for some years to entrust more and more to them.

In nothing done by either of these two wise fathers is more wisdom shown than in their sagacious, farseeing policy in regard to their sons. As they themselves had been taught to concentrate their energies upon useful occupation, for which society would pay as for value received, they had doubtless often conferred, and concluded that was the happiest and best life for their sons, instead of allowing them to fritter away the precious years of youth in aimless frivolity, to befollowed in later years by a disappointing and humiliating old age.

So the partnership of Boulton and Watt was renewed in the union of the sons. Gregory Watt's premature death four years later was such a blow to his father that some think he never was quite himself again. Gregory had displayed brilliant talents in the higher pursuits of science and literature, in which he took delight, and great things had been predicted from him. With the other two sons the business connection continued without change for forty years, until, when old men, they also retired like their fathers. They proved to be great managers, for notwithstanding the cessation of the patents which opened engine-building free to all, the business of the firm increased and became much more profitable than it had ever been before; indeed toward the close of the original partnership, and upon the triumph gained in the patent suits, the enterprise became so profitable as fully to satisfy the moderate desire of Watt, and to provide a sure source of income for his sons. This met all his wishes and removed the fears of becoming dependent that had so long haunted him.

The continued and increasing success of the Soho works was obviously owing to the new partners. They had some excellent assistants, but in the foremost place among all of them stands Murdoch, Watt's able, faithful and esteemed assistant for many years, who, both intellectually and in manly independence, was considered to exhibit no small resemblance to his revered master and friend. Never formally a partner in Soho (for he declined partnership as we have seen), he was placed on the footing of a partner by the sons in 1810, without risk, and received $5,000 per annum. From 1830 he lived in peaceful retirement and passed away in 1839. His remains were deposited in Handsworth Church near those of his friends and employers, Watt and Boulton (the one spot on earth he could have most desired). "A bust by Chantrey serves to perpetuate the remembrance of his manly and intelligent features, and of the mind of which these were a pleasing index." We may imagine the shades of Watt and Boulton, those friends so appropriately laid together, greeting their friend and employee: "Well done, thou good and faithful servant!" If ever there was one, Murdoch was the man, and Captain Jones his fellow.

We have referred to Watt's suggestion of the screw-propeller, and of the sketch of it sent to Dr. Small, September 30, 1770. The only record of any earlier suggestion of steam is that of Jonathan Hulls, in 1736, and which he set forth in a pamphlet entitled "A Description and Draught of a Newly Invented Machine for carrying vessels or ships out of or into any Harbour, Port or River, against Wind or Tide or in a Calm"; London, 1737. He described a large barge equipped with a Newcomen engine to be employed as a tug, fitted with fan (or paddle) wheels, towing a ship of war, but nothing further appears to have been done. Writing on this subject, Mr. Williamson says:

During his last visit to Greenock in 1816, Mr. Watt, in company with his friend, Mr. Walkinshaw—whom the author some years afterward heard relate the circumstance—made a voyage in a steamboat as far as Rothsay and back to Greenock—an excursion, which, in those days, occupied a greater portion of a whole day. Mr. Watt entered into conversation with the engineer of the boat, pointing out to him the method of "backing" the engine. With a footrule he demonstrated to him what was meant. Not succeeding, however, he at last, under the impulse of the ruling passion, threw off his overcoat, and, putting his hand to the engine himself, showed the practical application of his lecture. Previously to this, the "back-stroke" of the steamboat engine was either unknown, or not generally known. The practice was to stop the engine entirely a considerable time before the vessel reached the point of mooring, in order to allow for the gradual and natural diminution of her speed.

The naval review at Spithead, upon the close of the Crimean war in 1856, was the greatest up to that time. Ten vessels out of two hundred and fifty still had not steam power, but almost all the others were propelled by the screw—the spiral oar of Watt's letter of 1770—a red-letter day for the inventor.

Watt's early interest in locomotive steam-carriages, dating from Robison's having thrown out the idea to him, was never lost. On August 12, 1768, Dr. Small writes Watt, referring to the "peculiar improvements in them" the latter had made previous to that date. Seven months later he apprises Watt that "a patent formoving wheel-carriages by steam has been taken out by one Moore," adding "this comes of thy delays; do come to England with all possible speed." Watt replied "If linen-draper Moore does not use my engine to drive his chaises he can't drive them by steam." Here Watt hit the nail on the head; as with the steamship, so with the locomotive, his steam-engine was the indispensable power. In 1786 he states that he has a carriage model of some size in hand "and am resolved to try if God will work a miracle in favor of these carriages." Watt's doubt was based on the fact that they would take twenty pounds of coal and two cubic feet of water per horse-power on the common roads.

Another of Watt's recreations in his days of semi-retirement was the improvement of lamps. He wrote the famous inventor of the Argand burner fully upon the subject in August, 1787, and constructed some lamps which proved great successes.

The following year he invented an instrument for determining the specific gravities of liquids, which was generally adopted.

One of Watt's inventions was a new method of readily measuring distances by telescope, which he used in making his various surveys for canals. Such instruments are in general use to-day. Brough's treatise on "Mining" (10th ed., p. 228) gives a very complete account of them, and states that "the originalinstrument of this class is that invented by James Watt in 1771."

In his leisure hours, Watt invented an ingenious machine for drawing in perspective, using the double parallel ruler, then very little known and not at all used as far as Watt knew. Watt reports having made from fifty to eighty of these machines, which went to various parts of the world.

In 1810 Watt informs Berthollet that for several years he had felt unable, owing to the state of his health, to make chemical experiments. But idle he could not be; he must be at work upon something. As he often said, "Without a hobby-horse, what is life?" So the saying is reported, but we may conclude that the "horse" is here an interpolation, for the difference between "a horses" and "a hobby" is radical—a man can get off a horse.

Watt's next "hobby" fortunately became an engrossing occupation and kept him alert. This was a machine for copying sculpture. A machine he had seen in Paris for tracing and multiplying the dies of medals, suggested the other. After much labor and many experiments he did get some measure of success, and made a large head of Locke in yellow wood, and a small head of his friend Adam Smith.

Long did Watt toil at the new hobby in the garret where it had been created, but the garret proved too hot in summer and too cold in winter. March 14, 1810, he writes Berthollet and Levèque:

I still do a little in mechanics: a part of which, if I live to complete it, I shall have the honor of communicating to my friends in France.

He went steadily forward and succeeded in making some fine copies in 1814. For one of Sappho he gives dates and the hours required for various parts, making a total of thirty-nine. Some censorious Sabbatarians discovered that the day he was employed one hour "doing her breast with 1/8th drill" was Sabbath, which in one who belonged to a strict Scottish Covenanter family, betokened a sad fall from grace. When we consider that his health was then precarious, that he was debarred from chemical experiments, and depended solely upon mechanical subjects; that in all probability it was a stormy day (Sunday, February 3, 1811), knowing also that "Satan finds mischief still for idle hands to do," we hope our readers will pardon him for yielding to the irresistible temptation, even if on the holy Sabbath day for once he could not "get off" his captivating hobby.

The historical last workshop of the great worker with all its contents remains open to the public to-day just as it was when he passed away. Pilgrims from many lands visit it, as Shakespeare's birthplace, Burns' cottage, and Scott's Abbottsford attract their many thousands yearly. We recommend our readers to add to these this garret of Watt in their pilgrimages.

CHAPTER VIII

THE RECORD OF THE STEAM ENGINE

The Soho works, up to January, 1824, had completed 1164 steam engines, of a nominal horse-power of 25,945; from January, 1824, to 1854, 441 engines, and nominal horse-power, 25,278, making the total number 1605, of nominal horse-power, 51,223, and real horse-power, 167,319. Mulhall gives the total steam-power of the world as 50,150,000 horse-power in 1888. In 1880 it was only 34,150,000. Thus in eight years it increased, say, fifty per cent. Assuming the same rate of increase from 1888 to 1905, a similar period, it is to-day 75,000,000 nominal, which Engel says may be taken as one-half the effective power (vide Mulhall, "Steam," p. 546), the real horse-power in 1905 being 150,000,000. One horse-power raises ten tons a height of twelve inches per minute. Working eight hours, this is about 5,000 tons daily, or twelve times a man's work, and as the engine never tires, and can be run constantly, it follows that each horse-power it can exert equals thirty-six men's work; but, allowing for stoppages, let us say thirty men. The engines of a large ocean greyhound of 35,000 horse-power, running constantly from port to port, equal to three relays of twelve men per horse-power, is daily exerting the power of 1,260,000 men, or 105,000 horses. Assuming that all the steam engines in the world upon the average work double the hours of men, then the 150,000,000 horse-power in the world, each equal to two relays of twelve men per horse-power, exerts the power of 3,600,000,000 of men. There are only one-tenth as many male adults in the world, estimating one in five of the population.

If we assume that all steam engines work an average of only eight hours in the twenty-four, as men and horses do (those on duty longer hours are not under continuous exertion), it still follows that the 150,000,000 of effective steam-power, each doing the work of twelve men, equals the work of 1,800,000,000 of men, or of 150,000,000 of horses.

Engel estimated that in 1880 the value of world industries dependent upon steam was thirty-two thousand millions of dollars, and that in 1888 it had reached forty-three thousand millions of dollars. It is to-day doubtless more than sixty thousand millions of dollars, a great increase no doubt over 1880, but the one figure is as astounding as the other, for both mean nothing that can be grasped.

The chief steam-using countries are America, 14,400,000 horse-power in 1888; Britain, 9,200,000 horse-power nominal. If we add the British colonies and dependencies, 7,120,000 horse-power, the English-speaking race had three-fifths of all the steam-power of the world.

In 1840 Britain had only 620,000 horse-power nominal; the United States 760,000; the whole world had only 1,650,000 horse-power. To-day it has 75,000,000 nominal. So rapidly has steam extended its sway over most of the earth in less than the span of a man's life? There has never been any development in the world's history comparable to this, nor can we imagine that such a rapid transformation can ever come in the future. What the future is finally to bring forth even imagination is unable to conceive. No bounds can be set to its forthcoming possible, even probable, wonders, but as such a revolution as steam has brought must come from a superior force capable of displacing steam, this would necessarily be a much longer task than steam had in occupying an entirely new field without a rival.

The contrast between Newcomen and Watt is interesting. The Newcomen engine consumed twenty-eight pounds of coal per horse-power and made not exceeding three to four strokes per minute, the piston moving about fifty feet per minute. To-day, steam marine engines on one and one-third pounds of coal per horse-power—the monster ships using less—make from seventy to ninety revolutions per minute. "Destroyers" reach 400 per minute. Small steam engines, it is stated, have attained 600 revolutions per minute. The piston to-day is supposed to travel moderately when at 1,000 feet per minute, in a cylinder three feet long. This gives 166 revolutions per minute. With coal under the boilers costing one dollar per net ton, from say five pounds of coal for one cent there is one horse-power for three hours, or a day and a night of continuous running for eight cents.

Countless millions of men and of horses would be useless for the work of the steam-engine, for the seemingly miraculous quality steam possesses, that permits concentration, is as requisite as its expansive powers. One hundred thousand horse-power, or several hundred thousand horse-power, is placed under one roof and directed to the task required. Sixty-four thousand horse-power is concentrated in the hold of the great steamships now building. All this stupendous force is evolved, concentrated and regulated by science from the most unpromising of substances, cold water. Nothing man has discovered or imagined is to be named with the steam engine. It has no fellow. Franklin capturing the lightning, Morse annihilating space with the telegraph, Bell transmitting speech through the air by the telephone, are not less mysterious—being more ethereal, perhaps in one sense they are even more so—still, the labor of the world performed by heating cold water places Watt and his steam engine in a class apart by itself. Many are the inventions for applying power; his creates the power it applies.

Whether the steam engine has reached its climax, and gas, oil, or other agents are to be used extensively for power, in the near future, is a question now debated in scientific circles. Much progress has been made in using these substitutes, and more is probable, as one obstacle after another is overcome. Gas especially is coming forward, and oil is freely used. For reasons before stated, it seems to the writer that, where coal is plentiful, the day is distant when steam will not continue to be the principal source of power. It will be a world surpriser that beats one horse-power developed by one pound of coal. The power to do much more than this, however, lies theoretically in gas, but there come these wise words of Arago to mind: "Persons whose whole lives have been devoted to speculative labours are not aware how great the distance is between a scheme, apparently the best concerted, and its realisation." So true! Watt's ideas in the brain, and the steam engine that he had to evolve during nine long years, are somewhat akin to the great gulf between resolve and performance, the "good resolution" that soothes and the "act" that exalts.

The steam engine is Scotland's chief, tho not her only contribution to the material progress of the world. Watt was its inventor, we might almost write Creator, and so multiform were the successive steps. Symington by the steamship stretched one arm of it over the water; Stephenson by the locomotive stretched the other over the land. Thus was the world brought under its sway and conditions of human life transformed? Watt and Symington were born in Scotland within a few miles of each other. Stephenson's forbears moved from Scotland south of the line previous to his birth, as Fulton's parents removed from Scotland to America, so that both Stephenson and Fulton could boast with Gladstone that the blood in their veins was Scotch.

The history of the world has no parallel to the change effected by the inventions of these three men. Strange that little Scotland, with only 1,500,000 people, in 1791, about one-half the population of New York City, should have been the mother of such a triad, and that her second "mighty three" (Wallace, Bruce and Burns always first), should have been of the same generation, working upon the earth near each other at the same time. The Watt engine appeared in 1782; the steamship in 1801; the locomotive thirteen years later, in 1814. Thus thirty-two years after its appearance Watt's steam-engine had conquered both sea and land.

The sociologist may theorise, but plain people will remember that men do not gather grapes from thorns, nor figs from thistles. There must be something in the soil which produces such men; something in the poverty that compels exertion; something in the "land of the mountain and the flood" that stirs the imagination; something in the history of centuries of struggle for national and spiritual independence; much in the system of compulsory and universal free education; something of all these elements mingling in the blood that tells, and enables Scotland to contribute so largely to the progress of the world.

Strange reticence is shown by all Watt's historians regarding his religious and political views. Williamson, the earliest author of his memoirs, is full of interesting facts obtained from people in Greenock who had known Watt well. The hesitation shown by him as to Watt's orthodoxy in his otherwise highly eulogistic tribute, attracts attention. He says:

We could desire to know more of the state of those affections which are more purely spiritual by their nature and origin—his disposition to those supreme truths of Revelation, which alone really elevate and purify the soul. In the absence of much information of a very positive kind in regard to such points of character and life, we instinctively revert in a case like this to the principles and maxims of an infantile and early training. Remembering the piety portrayed in the ancestors of this great man, one cannot but cling to the hope that his many virtues reposed on a substratum of more than merely moral excellence. Let us cherish the hope that the calm which rested on the spirit of the pilgrim ... was one that caught its radiance from a far higher sphere than that of the purest human philosophy.

Watt's breaking of the Sabbath before recorded must have seemed to that stern Calvinist a heinous sin, justifying grave doubts of Watt's spiritual condition, his "moral excellence" to the contrary notwithstanding. Williamson's estimate of moral excellence had recently been described by Burns:

But then, nae thanks to him for a' that,Nae godly symptom ye can ca' that,It's naething but a milder featureOf our poor sinfu' corrupt nature.Ye'll get the best o' moral works,Many black gentoos and pagan works,Or hunters wild on PonotaxiWha never heard of orthodoxy.

Williamson's doubts had much stronger foundation in Watt's non-attendance at church, for, as we shall see from his letter to DeLuc, July, 1788, he had never attended the "meeting-house" (dissenting church) in Birmingham altho he claimed to be still a member of the Presbyterian body in declining the sheriffalty.

It seems probable that Watt, in his theological views, like Priestley and others of the Lunar Society, was in advance of his age, and more or less in accord with Burns, who was then astonishing his countrymen. Perhaps he had forstalled Dean Stanley's advice in his rectorial address to the students of St. Andrew's University: "go to Burns for your theology," yet he remained a deeply religious man to the end, as we see from his letter (page 216), at the age of seventy-six.

We know that politically Watt was in advance of his times for the prime minister pronounced him "a sad radical." He was with Burns politically at all events. Watt's eldest son, then in Paris, was carried away by the French Revolution, and Muirhead suggests that the prime minister must have confounded father and son, but it seems unreasonable to suppose that he could have been so misled as to mistake the doings of the famous Watt in Birmingham for those of his impulsive son in France.

The French Revolution exerted a powerful influence in Britain, especially in the north of England and south of Scotland, which have much in common. The Lunar Society of Birmingham was intensely interested. At one of the meetings in the summer of 1788, held at her father's house, Mrs. Schimmelpenniack records that Mr. Boulton presented to the company his son, just returned from a long sojourn in Paris, who gave a vivid account of proceedings there, Watt and Dr. Priestly being present. A few months later the revolution broke out. Young Harry Priestley, a son of the Doctor's, one evening burst into the drawing-room, waving his hat and crying, "Hurrah! Liberty, Reason, Brotherly Love forever! Down with kingcraft and priestcraft! The majesty of the people forever! France is free!" Dr. Priestley was deeply stirred and became the most prominent of all in the cause of the rights of man. He hailed the acts of the National Assembly abolishing monarchy, nobility and church. He was often engaged in discussions with the local clergy on theological dogmas. He wrote a pamphlet upon the French Revolution, and Burke attacked him in the House of Commons. All this naturally concentrated local opposition upon him as leader. The enthusiasts mistakenly determined to have a public dinner to celebrate the anniversary of the Revolution, and no less than eighty gentlemen attended, altho many advised against it. Priestley himself was not present. A mob collected outside and demolished the windows. The cry was raised, "To the new meeting-house!" the chapel in which Priestley ministered. The chapel was set on fire. Thence the riot proceeded to Priestley's house. The doctor and his family, being warned, had left shortly before. The house was at the mercy of the mob, which broke in, destroyed furniture, chemical laboratory and library, and finally set fire to the house. Some of the very best citizens suffered in like manner. Mr. Ryland, one of the most munificent benefactors of the town, Mr. Taylor, the banker, and Hutton, the estimable book-seller, were among the number. The home of Dr. Withering, member of the Lunar Society, was entered, but the timely arrival of troops saved it from destruction. The members of the Lunar Society, or the "lunatics," as they were popularly called, were especially marked for attack. The mob cried, "No philosophers!" "Church and King forever!" All this put Boulton and Watt upon their guard, for they were prominent members of the society. They called their workmen together, explained the criminally of the rioters, and placed arms in their hands on their promise to defend them if attacked. Meanwhile everything portable was packed up ready to be removed.

Watt wrote to Mr. DeLuc, July 19, 1791:

Though our principles, which are well known, as friends to the established government and enemies of republican principles, should have been our protection from a mob whose watchword was Church and King, yet our safety was principally owing to most of the Dissenters living south of the town; for after the first moment they did not seem over-nice in their discrimination of religion and principles. I, among others, was pointed out as a Presbyterian, though I never was in a meeting-house (Dissenting Church) in Birmingham, and Mr. Boulton is well-known as a Churchman. We had everything most portable packed up, fearing the worst. However, all is well with us.

From all this we gather the impression that Radical principles had permeated the leading minds of Birmingham to a considerable extent, probably around the Lunar Society district in greater measure than in other quarters, altho clubs of ardent supporters were formed in London and the principal provincial cities.

In the political field, we have only one appearance of Watt reported. Early in 1784, we find him taking the lead in getting up a loyal address to the king on the appointment as prime minister of Pitt, who proposed to tax coal, iron, copper and other raw materials of manufacture to the amount of $5,000,000 per year, a considerable sum in those days when manufacturing was in its infancy. Boulton also joined in opposition. They wisely held that for a manufacturing nation "to tax raw materials was suicidal: let taxes be laid upon luxuries, upon vices, and, if you like, upon property; tax riches when got, but not the means of getting them. Of all things don't cut open the hen that lays the golden eggs."

Watt's services were enlisted and he drew up a paper for circulation upon the subject. The policy failed, and soon after Pitt was converted to sounder doctrines by Adam Smith's "Wealth of Nations." Free trade has ruled Britain ever since, and, being the country that could manufacture cheapest, and indeed, the only manufacturing country for many years, this policy has made her the richest, per capita, of all nations. The day may be not far distant when America, soon to be the cheapest manufacturing country for many, as it already is for a few, staple articles, will be crying for free trade, and urging free entrance to the markets of the world. To tax the luxuries and vices, to tax wealth got and not in the making, as proposed by Watt and Boulton, is the policy to follow. Watt shows himself to have been a profound economist.

Watt had cause for deep anxiety for his eldest son, James, who had taken an active part in the agitation. He and his friend, Mr. Cooper of Manchester, were appointed deputies by the "Constitutional Society," to proceed to Paris and present an address of congratulation to the Jacobin Club. Young Watt was carried away, and became intimate with the leaders. Southey says he actually prevented a duel between Danton and Robespierre by appearing on the ground and remonstrating with them, pointing out that if either fell the cause must suffer.

Upon young Watt's return, king's messengers arrived in Birmingham and seized persons concerned in seditious correspondence. Watt suggests that Boulton should see his son and arrange for his leaving for America, or some foreign land, for a time. This proved to be unnecessary; his son was not arrested, and in a short time all was forgotten. He entered the works with Boulton's son as partner, and became an admirable manager. To-day we regard his mild republicanism, his alliance with Jacobin leaders, and especially his bold intervention in the quarrel between two of the principal actors in the tragedy of the French Revolution, as "a ribbon in the cap of youth." That his douce father did the same and was proud of his eldest born seems probable. Our readers will also judge for themselves whether the proud father had not himself a strong liking for democratic principles, "the rights of the people," "the royalty of man," which Burns was then blazing forth, and held such sentiments as quite justified the prime minister's accusation that he was "a sad radical."

In Britain, since Watt's day, all traces of opposition to monarchy aroused by the French Revolution have disappeared, as completely as the monarchy of King George. The "limited monarchy" of to-day, developed during the admirable reign of Queen Victoria, has taken its place. The French abolished monarchy by a frontal attack upon the citadel, involving serious loss. Not such the policy of the colder Briton. He won his great victory, losing nothing, by flanking the position. That the king "could do no wrong," is a doctrine almost coeval with modern history, flowing from the "divine right" of kings, and, as such, was quietly accepted. It needed only to be properly harnessed to become a very serviceable agent for registering the people's will.

It was obvious that the acceptance of the doctrine that the king could do no wrong involved the duty of proving the truth of the axiom, and it was equally obvious that the only possible way of doing this was that the king should not be allowed to do anything. Hence he was made the mouthpiece of his ministers, and it is not the king, but they, who, being fallible men, may occasionally err. The monarch, in losing power to do anything has gained power to influence everything. The ministers hold office through the approval of the House of Commons. Members of that house are elected by the people. Thus stands government in Britain "broad-based upon the people's will."

All that the revolutionists of Watt's day desired has, in substance, been obtained, and Britain has become in truth a "crowned republic," with "government of the people, for the people, and by the people." This steady and beneficent development was peaceably attained. The difference between the French and British methods is that between revolution and evolution.

In America's political domain, a similar evolution has been even more silently at work than in Britain during the past century, and is not yet exhausted—the transformation of a loose confederacy of sovereign states, with different laws, into one solid government, which assumes control and insures uniformity over one department after another. The centripetal forces grow stronger with the years; power leaves the individual states and drifts to Washington, as the necessity for each successive change becomes apparent. In the regulation of interstate commerce, of trusts, and in other fields, final authority over the whole land gravitates more and more to Washington. It is a beneficent movement, likely to result in uniform national laws upon many subjects in which present diversity creates confusion. Marriage and divorce laws, bankruptcy laws, corporation charter privileges, and many other important questions may be expected to become uniform under this evolutionary process. The Supreme Court decision that the Union was an indissoluble union of indissoluble states, carries with it finally uniform regulation of many interstate problems, in every respect salutary, and indispensable for the perfect union of the American people.

CHAPTER IX

WATT IN OLD AGE

Watt gracefully glided into old age. This is the great test of success in life. To every stage a laurel, but to happy old age the crown. It was different with his friend Boulton, who continued to frequent the works and busy himself in affairs much as before, altho approaching his eightieth year. Watt could still occupy himself in his garret, where his "mind to him a Kingdom was," upon the scientific pursuits which charmed him. He revisited Paris in 1802 and renewed acquaintances with his old friends, with whom he spent five weeks. He frequently treated himself to tours throughout England, Scotland and Wales. In the latter country, he purchased a property which attracted him by its beauties, and which he greatly improved. It became at a later date, under his son, quite an extensive estate, much diversified, and not lacking altogether the stern grandeur of his native Scotland. He planted trees and took intense delight in his garden, being very fond of flowers. The farmhouse gave him a comfortable home upon his visits. The fine woods which now richly clothe the valley and agreeablydiversify the river and mountain scenery were chiefly planted under his superintendence, many by his own hand. In short, the blood in his veins, the lessons of his childhood that made him a "child of the mist," happy in roaming among the hills, reasserted their power in old age as the Celtic element powerfully does. He turned more and more to nature.

"That never yet betrayed the heart that loved her—"

We see him strolling through his woods, and imagine him crooning to himself from that marvellous memory that forgot no gem:

For I have learnedTo look on nature, not as in the hourOf thoughtless youth; but hearing oftentimesThe still, sad music of humanity,Nor harsh, nor grating, though of ample powerTo chasten and subdue. And I have feltA presence that disturbs me with the joyOf elevated thoughts; a sense sublimeOf something far more deeply interfused, Whose dwelling is the light of setting suns,And the round ocean and the living air,And the blue sky, and in the mind of man:A motion and a spirit, that impelsAll thinking things, all objects of all thought,And rolls through all things. Therefore am I stillA lover of the meadows and the woods,And mountains; and of all that we beholdFrom this green earth.

Twice Watt was requested to undertake the honor of the shrievalty; in 1803 that of Staffordshire, and in 1816 that of Radnorshire, both of which were positively declined.

He finally found it necessary to declare that he was not a member of the Church of England, but of the Presbyterian church of Scotland, a reason which in that day was conclusive.

In 1816, he was in his eighty-first year, and no difficulty seems then to have been found for excusing him, for it seems the assumption of the duties was compulsory. It was "the voice of age resistless in its feebleness."

The day had come when Watt awakened to one of the saddest of all truths, that his friends were one by one rapidly passing away, the circle ever narrowing, the few whose places never could be filled becoming fewer, he in the centre left more and more alone. Nothing grieved Watt so much as this. In 1794 his partner, Roebuck, fell; in 1799, his inseparable friend, and supporter in his hour of need, Dr. Black, and also Withering of the Lunar Society; and in 1802 Darwin "of the silver song," one of his earliest English friends. In 1804, his brilliant son Gregory died, a terrible shock. In 1805, his first Glasgow College intimate, Robison; Dr. Beddoes in 1808; Boulton, his partner, in 1809; Dr. Wilson in 1811; DeLuc in 1817. Many other friends of less distinction fell in these years who were not less dear to him. He says, "by one friend's withdrawing after another," he felt himself "in danger of standing alone among strangers, the son of later times."

He writes to Boulton on November 23, 1802:

We cannot help feeling, with deep regret, the circle of our old friends gradually diminishing, while our ability to increase it by new ones is equally diminished; but perhaps it is a wise dispensation of Providence so to diminish our enjoyments in this world, that when our turn comes we may leave it without regret.

He writes to another correspondent, July 12, 1810:

I, in particular, have reason to thank God that he has preserved me so well as I am, to so late a period, while the greater part of my contemporaries, healthier and younger men, have passed "the bourne from which no traveller returns." It is, however, a painful contemplation to see so many who were dear to us pass away before us; and our consolation should be, that as Providence has been pleased to prolong our life, we should render ourselves as useful to society as we can while we live.

And again, when seventy-six years of age, January, 1812, he writes:

On these subjects I can offer no other consolations than what are derived from religion: they have only gone before us a little while, in that path we all must tread, and we should be thankful they were spared so long to their friends and the world.

Sir Walter Scott declares:

That is the worst part of life when its earlier path is trod. If my limbs get stiff, my walks are made shorter, and my rides slower; if my eyes fail me, I can use glasses and a large print: if I get a little deaf, I comfort myself that except in a few instances I shall be no great loser by missing one full half of what is spoken: *but I feel the loneliness of age when my companions and friends are taken from me.*

All his life until retiring from business, Watt's care was to obtain sufficient for the support of himself and family upon the most modest scale. He had no surplus to devote to ends beyond self, but as soon as he retired with a small competence it was different, and we accordingly find him promptly beginning to apply some portion of his still small revenue to philanthropical ends. Naturally, his thoughts reverted first to his native town and the university to which he owed so much.

In 1808 he founded the Watt Prize in Glasgow University, saying:

Entertaining a due sense of the many favours conferred upon me by the University of Glasgow, I wish to leave them some memorial of my gratitude, and, at the same time, to excite a spirit of inquiry and exertion among the students of Natural Philosophy and Chemistry attending the College; which appears to me the more useful, as the very existence of Britain, as a nation, seems to me, in great measure, to depend upon her exertions in science and in the arts.

The University conferred the degree of LL.D. upon him in 1774, and its great engineering laboratory bears his name.

In 1816, he made a donation to the town of Greenock for scientific books, stating it to be his intention to form the beginning of a scientific library for the instruction of the youth of Greenock, in the hope of prompting others to add to it, and of rendering his townsmen as eminent for their knowledge as they are for the spirit of enterprise.

This has grown to be a library containing 15,000 volumes, and is a valuable adjunct of the Watt Institution, founded by his son in memory of his father, which is to-day the educational centre of Greenock. Its entrance is adorned by a remarkably fine statue of Watt, funds for which were raised by public subscription.

Many societies honored the great inventor. He was a fellow of the Royal Society of Edinburgh, the Royal Society of London, Member of the Batavian Society, correspondent of the French Academy of Sciences, and was one of the eight Foreign Associates of the French Academy of Sciences.

Watt's almost morbid dislike for publicity leaves many well-known acts of kindness and charity hidden from all save the recipients. Muirhead assures us that such gifts as we can well believe were not wanting. Watt's character as a kindly neighbor always stood high. He was one of those "who will not receive a reward for that for which God accounts Himself a debtor—persons that dare trust God with their charity, and without a witness."

In the autumn of 1819 an illness of no great apparent severity caused some little anxiety to Watt's family, and was soon recognised by himself as the messenger sent to apprise him of his end. This summons he met with the calm and tranquil mind, that, looking backward, could have found little of serious nature to repent, and looking forward, found nothing to fear. "He often expressed his gratitude to the Giver of All Good who had so signally prospered the work of his hands and blessed him with length of days and riches and honour." On August 19, 1819, aged 83, in his own home at Heathfield, he tranquillybreathed his last, deeply mourned by all who were privileged to know him. In the parish churchyard, alongside of Boulton, he was most appropriately laid to rest. Thus the two strong men, lifelong friends and partners, who had never had a serious difference, "lovely and pleasant in their lives, in their death were not divided."

It may be doubted whether there be on record so charming a business connection as that of Boulton and Watt; in their own increasingly close union for twenty-five years, and, at its expiration, in the renewal of that union in their sons under the same title; in their sons' close union as friends without friction as in the first generation; in the wonderful progress of the world resulting from their works; in their lying down side by side in death upon the bosom of Mother Earth in the quiet churchyard, as they had stood side by side in the battle of life; and in the faithful servant Murdoch joining them at the last, as he had joined them in his prime. In the sweet and precious influences which emanate from all this, may we not gratefully make acknowledgment that in contemplation thereof we are lifted into a higher atmosphere, refreshed, encouraged, and bettered by the true story of men like ourselves, whom if we can never hope to equal, we may at least try in part to imitate.

A meeting was called in London to take steps for a monument to Watt to be placed in Westminster Abbey. The prime minister presided and announced a subscription of five hundred pounds sterling from His Majesty. It may truly be said that

A meeting more distinguished by rank, station and talent, was never before assembled to do honour to genius, and to modest and retiring worth; and a more spontaneous, noble, and discriminating testimony was never borne to the virtues, talents, and public services of any individual, in any age or country.

The result was the colossal statue by Chantrey which bears the following inscription, pronounced to be beyond comparison "the finest lapidary inscription in the English language." It is from the pen of Lord Brougham:

NOT TO PERPETUATE A NAME
WHICH MUST ENDURE WHILE THE PEACEFUL ARTS FLOURISH
BUT TO SHEW
THAT MANKIND HAVE LEARNT TO HONOUR THOSE
WHO BEST DESERVE THEIR GRATITUDE
THE KING
HIS MINISTERS, AND MANY OF THE NOBLES
AND COMMONERS OF THE REALM
RAISED THIS MONUMENT TO
JAMES WATT
WHO DIRECTING THE FORCE OF AN ORIGINAL GENIUS
EARLY EXERCISED IN PHILOSOPHIC RESEARCH
TO THE IMPROVEMENT OF
THE STEAM-ENGINE
ENLARGED THE RESOURCES OF HIS COUNTRY
INCREASED THE POWER OF MAN
AND ROSE TO AN EMINENT PLACE

AMONG THE MOST ILLUSTRIOUS FOLLOWERS OF SCIENCE
AND THE REAL BENEFACTORS OF THE WORLD
BORN AT GREENOCK MDCCXXXVI
DIED AT HEATHFIELD IN STAFFORDSHIRE MDCCCXIX

CHAPTER X

Watt, the Inventor and Discoverer

In the foregoing pages an effort has been made to follow and describe Watt's work in detail as it was performed, but we believe our readers will thank us for presenting the opinions of a few of the highest scientific and legal authorities upon what Watt really did. Lord Brougham has this to say of Watt:

One of the most astonishing circumstances in this truly great man was the versatility of his talents. His accomplishments were so various, the powers of his mind were so vast, and yet of such universal application, that it was hard to say whether we should most admire the extraordinary grasp of his understanding, or the accuracy of nice research with which he could bring it to bear upon the most minute objects of investigation. I forget of whom it was said, that his mind resembled the trunk of an elephant, which can pick up straws and tear up trees by the roots. Mr. Watt in some sort resembled the greatest and most celebrated of his own inventions; of which we are at a loss whether most to wonder at the power of grappling with the mightiest objects, or of handling the most minute; so that while nothing seems too large for its grasp, nothing seems too small for the delicacy of its touch; which can cleave rocks and pour forth rivers from the bowels of the earth, and with perfect exactness, though not with greater ease, fashion the head of a pin, or strike the impress of some curious die. Now those who knew Mr. Watt, had to contemplate a man whose genius could create such an engine, and indulge in the most abstruse speculations of philosophy, and could at once pass from the most sublime researches of geology and physical astronomy, the formation of our globe, and the structure of the universe, to the manufacture of a needle or a nail; who could discuss in the same conversation, and with equal accuracy, if not with the same consummate skill, the most forbidding details of art, and the elegances of classical literature; the most abstruse branches of science, and the niceties of verbal criticism.

There was one quality in Mr. Watt which most honorably distinguished him from too many inventors, and was worthy of all imitation; he was not only entirely free from jealousy, but he exercised a careful and scrupulous self-denial, and was anxious not to appear, even by accident, as appropriating to himself that which he thought belonged in part to others. I have heard him refuse the honor universally ascribed to him, of being inventor of the steam-engine, and call himself simply its improver; though, in my mind, to doubt his right to that honor would be as inaccurate as to question Sir Isaac Newton's claim to his greatest discoveries, because Descartes in mathematics, and Galileo in astronomy and mechanics, had preceded him; or to deny the merits of his illustrious successor, because galvanism was not his discovery, though before his time it had remained as useless to science as the instrument called a steam-engine was to the arts before Mr. Watt. The only jealousy I have known him betray was with respect to others, in the nice adjustment he was fond of giving to the claims of inventors. Justly prizing scientific discovery above all other possessions, he deemed the title to it so sacred, that you might hear him arguing by the hour to settle disputed rights; and if you ever perceived his temper ruffled, it was when one man's invention was claimed by, or given to, another; or when a clumsy adulation pressed upon himself that which he knew to be not his own.

Sir Humphrey Davy says:

I consider it as a duty incumbent on me to endeavor to set forth his peculiar and exalted merits, which live in the recollection of his contemporaries and will transmit his name with immortal glory to posterity. Those who consider James Watt only as a great practical mechanic form a very erroneous idea of his character; he was equally distinguished as a natural philosopher and a chemist, and his inventions demonstrate his profound knowledge of those sciences, and that peculiar characteristic of genius, the union of them for practical application. The steam engine before his time was a rude machine, the result of simple experiments on the compression of the atmosphere, and the condensation of steam. Mr. Watt's improvements were not produced by accidental circumstances or by a single ingenious thought; they were founded on delicate and refined experiments, connected with the discoveries of Dr. Black. He had to investigate the cause of the cold produced by evaporation, of the heat occasioned by the condensation of steam—to determine the source of the air appearing when water was acted upon by an exhausting power; the ratio of the volume of steam to its generating water, and the law by which the elasticity of steam increased with the temperature; labor, time, numerous and difficult experiments, were required for the ultimate result; and when his principle was obtained, the application of it to produce the movement of machinery demanded a new species of intellectual and experimental labor.

The Archimedes of the ancient world by his mechanical inventions arrested the course of the Romans, and stayed for a time the downfall of his country. How much more has our modern Archimedes done? He has permanently elevated the strength and wealth of his great empire: and, during the last long war, his inventions; and their application were amongst the great means which enabled Britain to display power and resources so infinitely above what might have been expected from the numerical strength of her population. Archimedes valued principally abstract science; James Watt, on the contrary, brought every principle to some practical use; and, as it were, made science descend from heaven to earth. The great inventions of the Syracusan died with him—those of our philosopher live, and their utility and importance are daily more felt; they are among the grand results which place civilised above savage man—which secure the triumph of intellect, and exalt genius and moral force over mere brutal strength, courage and numbers.

Sir James Mackintosh says:

It may be presumptuous in me to add anything in my own words to such just and exalted praise. Let me rather borrow the language in which the great father of modern philosophy, Lord Bacon himself, has spoken of inventors in the arts of life. In a beautiful, though not very generally read fragment of his, called the New Atlantis, a voyage to an imaginary island, he has imagined a university, or rather royal society, under the name of Solomon's House, or the College of the Six Days' Works; and among the various buildings appropriated to this institution, he describes a gallery destined to contain the statues of inventors. He does not disdain to place in it not only the inventor of one of the greatest instruments of science, but the discoverer of the use of the silkworm, and of other still more humble contrivances for the comfort of man. What place would Lord Bacon have assigned in such a gallery to the statue of Mr. Watt? Is it too much to say, that, considering the magnitude of the discoveries, the genius and science necessary to make them, and the benefits arising from them to the world, that statue must have been placed at the head of those of all inventors in all ages and nations. In another part of his writings the same great man illustrates the dignity of useful inventions by one of those happy allusions to the beautiful mythology of the ancients, which he often employs to illuminate as well as to decorate reason. "The dignity," says he, "of this end of endowment of man's life with new commodity appeareth, by the estimation that antiquity made of such as guided thereunto; for whereas founders of states, lawgivers, extirpators of tyrants, fathers of the people, were honored but with the titles of demigods, inventors were ever consecrated amongst the gods themselves."

The Earl of Aberdeen says:

It would ill become me to attempt to add to the eulogy which you have already heard on the distinguished individual whose genius and talents we have met this day to acknowledge. That eulogy has been pronounced by those whose praises are well calculated to confer honor, even upon him whose name does honor to his country. I feel in common with them, although I can but ill express that intense admiration which the bare recollection of those discoveries must excite, which have rendered us familiar with a power before nearly unknown, and which have taught us to wield, almost at will, perhaps the mightiest instrument ever intrusted to the hands of man. I feel, too, that in erecting a monument to his memory, placed, as it may be, among the memorials of kings, and heroes, and statesmen, and philosophers, that it will be then in its proper place; and most in its proper place, if in the midst of those who have been most distinguished by their usefulness to mankind, and by the spotless integrity of their lives.

Lord Jeffrey says:

This name fortunately needs no commemoration of ours; for he that bore it survived to see it crowned with undisputed and unenvied honors; and many generations will probably pass away, before it shall have gathered "all its fame." We have said that Mr. Watt was the great *improver* of the steam engine; but, in truth, as to all that is admirable in its structure, or vast in its utility, he should rather be described as its *inventor*. It was by his inventions that its action was so regulated, as to make it capable of being applied to the finest and most delicate manufactures, and its power so increased, as to set weight and solidity at defiance. By his admirable contrivance, it has become a thing stupendous alike for its force and its flexibility, for the prodigious power which it can exert, and the ease, and precision, and ductility, with which it can be varied, distributed, and applied. The trunk of an elephant, that can pick up a pin or rend an oak, is as nothing to it. It can engrave a seal, and crush masses of obdurate metal before it; draw out, without breaking, a thread as fine as gossamer, and lift a ship of war like a bauble in the air. It can embroider muslin and forge anchors, cut steel into ribbons, and impel loaded vessels against the fury of the winds and waves.

It would be difficult to estimate the value of the benefits which these inventions have conferred upon this country. There is no branch of industry that has not been indebted to them; and, in all the most material, they have not only widened most magnificently the field of its exertions, but multiplied a thousandfold the amount of its productions. It is our improved steam engine that has fought the battles of Europe, and exalted and sustained, through the late tremendous contest, the political greatness of our land. It is the same great power which now enables us to pay the interest of our debt, and to maintain the arduous struggle in which we are still engaged (1819), with the skill and capital of countries less oppressed with taxation. But these are poor and narrow views of its importance. It has increased indefinitely the mass of human comforts and enjoyments, and rendered cheap and accessible, all over the world, the materials of wealth and prosperity. It has armed the feeble hand of man, in short, with a power to which no limits can be assigned; completed the dominion of mind over the most refractory qualities of matter; and laid a sure foundation for all those future miracles of mechanical power which are to aid and reward the labors of after generations. It is to the genius of one man, too, that all this is mainly owing; and certainly no man ever bestowed such a gift on his kind. The blessing is not only universal, but unbounded; and the fabled inventors of the plough and the loom, who were deified by the erring gratitude of their rude contemporaries, conferred less important benefits on mankind than the inventor of our present steam engine.

This will be the fame of Watt with future generations; and it is sufficient for his race and his country. But to those to whom he more immediately belonged, who lived in his society and enjoyed his conversation, it is not, perhaps, the character in which he will be most frequently recalled—most deeply lamented—or even most highly admired.

We shall end by quoting the greatest living authority, Lord Kelvin, now Lord Chancellor of Glasgow University, which Watt and he have done so much to render famous:

Precisely that single-acting, high-pressure, syringe-engine, made and experimented on by James Watt one hundred and forty years ago in his Glasgow College workshop, now in 1901, with the addition of a surface-condenser cooled by air to receive the waste steam, and a pump to return the water thence to the boiler, constitutes the common-road motor, which, in the opinion of many good judges, is the most successful of all the different motors which have been made and tried within the last few years. Without a condenser, Watt's high-pressure, single-acting engine of 1761, only needs the cylinder-cover with piston-rod passing steam-tight through it (as introduced by Watt himself in subsequent developments), and the valves proper for admitting steam on both sides of the piston and for working expansively, to make it the very engine, which, during the whole of the past century, has done practically all the steam work of the world, and is doing it still, except on the sea or lakes or rivers, where there is plenty of condensing water. Even the double and triple and quadruple expansion engines, by which the highest modern economy for power and steam engines has been obtained, are splendid mechanical developments of the principle of expansion, discovered and published by Watt, and used, though to a comparatively limited extent, in his own engines.

Thus during the five years from 1761-66 Watt had worked out all the principles and invented all that was essential in the details for realising them in the most perfect steam engines of the present day.

So passes Watt from view as the discoverer and inventor of the "most powerful instrument in the hands of man to alter the face of the physical world." He takes his place "at the head of all inventors of all ages and all nations."

CHAPTER XI

WATT, THE MAN

Of Watt, the genius, possessed of abilities far beyond those of other men, a scientist and philosopher, a mechanician and a craftsman, one who gravitated without effort to the top of every society, and who, even when a young workman, made his workshop the meeting-place of the leaders of Glasgow University for the interchange of views upon the highest and most abstruse subjects—with all this we have already dealt, but it is only part, and not the nobler part. He excelled all his fellows in knowledge, but there is much beyond mere knowledge in man. Strip Watt of all those commanding talents that brought him primacy without effort, for no man ever avoided precedence more persistently than he, and the question still remains: what manner of man was he, as man? Surely our readers would esteem the task but half done that revealed only what was unusual in Watt's head. What of his heart? is naturally asked. We hasten to record that in the domain of the personal graces and virtues, we have evidence of his excellence as copious and assured as for his pre-eminence in invention and discovery.

We cite the testimony of those who knew him best. It is seldom that a great man is so fortunate in his eulogists. The picture drawn of him by his friend, Lord Jeffrey, must rank as one of the finest ever produced, as portrait and tribute combined. That it is a discriminating statement, altho so eulogistic, may well be accepted, since numerous contributory proofs are given by others of Watt's personal characteristics. Says Lord Jeffrey:

Independently of his great attainments in mechanics, Mr. Watt was an extraordinary, and in many respects a wonderful man. Perhaps no individual in his age possessed so much and such varied and exact information—had read so much, or remembered what he had read so accurately and well. He had infinite quickness of apprehension, a prodigious memory, and a certain rectifying and methodising power of understanding, which extracted something precious out of all that was presented to it. His stores of miscellaneous knowledge were immense, and yet less astonishing than the command he had at all times over them. It seemed as if every subject that was casually started in conversation with him, had been that which he had been last occupied in studying and exhausting; such was the copiousness, the precision, and the admirable clearness of the information which he poured out upon it without effort or hesitation. Nor was this promptitude and compass of knowledge confined in any degree to the studies connected with his ordinary pursuits. That he should have been minutely and extensively skilled in chemistry and the arts, and in most of the branches of physical science, might perhaps have been conjectured; but it could not have been inferred from his usual occupations, and probably is not generally known, that he was curiously learned in many branches of antiquity, metaphysics, medicine, and etymology, and perfectly at home in all the details of architecture, music and law. He was well acquainted, too, with most of the modern languages, and familiar with their most recent literature. Nor was it at all extraordinary to hear the great mechanician and engineer detailing and expounding, for hours together, the metaphysical theories of the German logicians, or criticising the measures or the matter of the German poetry.

His astonishing memory was aided, no doubt, in a great measure, by a still higher and rarer faculty—by his power of digesting and arranging in its proper place all the information he received, and of casting aside and rejecting, as it were instinctively, whatever was worthless or immaterial. Every conception that was suggested to his mind seemed instantly to take its place among its other rich furniture, and to be condensed into the smallest and most convenient form. He never appeared, therefore, to be at all encumbered or perplexed with the *verbiage* of the dull books he perused, or the idle talk to which he listened; but to have at once extracted, by a kind of intellectual alchemy, all that was worthy of attention, and to have reduced it, for his own use, to its true value and to its simplest form. And thus it often happened that a great deal more was learned from his brief and vigorous account of the theories and arguments of tedious writers, than an ordinary student could ever have derived from the most painful study of the originals, and that errors and absurdities became manifest from the mere clearness and plainness of his statement of them, which might have deluded and perplexed most of his hearers without that invaluable assistance.

It is needless to say, that, with those vast resources, his conversation was at all times rich and instructive in no ordinary degree; but it was, if possible, still more pleasing than wise, and had all the charms of familiarity, with all the substantial treasures of knowledge. No man could be more social in his spirit, less assuming or fastidious in his manners, or more kind and indulgent toward all who approached him. He rather liked to talk, at least in his latter years, but though he took a considerable share of the conversation, he rarely suggested the topics on which it was to turn, but readily and quietly took up whatever was presented by those around him, and astonished the idle and barren propounders of an ordinary theme, by the treasures which he drew from the mine they had inconsciously opened. He generally seemed, indeed, to have no choice or predilection for one subject of discourse rather than another; but allowed his mind, like a great cyclopædia, to be opened at any letter his associates might choose to turn up, and only endeavour to select, from his inexhaustible stores, what might be best adapted to the taste of his present hearers. As to their capacity he gave himself no trouble; and, indeed, such was his singular talent for making all things plain, clear, and intelligible, that scarcely any one could be aware of such a deficiency in his presence. His talk, too, though overflowing with information, had no resemblance to lecturing or solemn discoursing, but, on the contrary, was full of colloquial spirit and pleasantry. He had a certain quiet and grave humour, which ran through most of his conversation, and a vein of temperate jocularity, which gave infinite zest and effect to the condensed and inexhaustible information which formed its main staple and characteristic. There was a little air of affected testiness, and a tone of pretended rebuke and contradiction, with which he used to address his younger friends, that was always felt by them as an endearing mark of his kindness and familiarity, and prized accordingly, far beyond all the solemn compliments that ever proceeded from the lips of authority. His voice was deep and powerful, although he commonly spoke in a low and somewhat monotonous tone, which harmonised admirably with the weight and brevity of his observations, and set off to the greatest advantage the pleasant anecdotes, which he delivered with the same grave brow, and the same calm smile playing soberly on his lips. There was nothing of effort indeed, or impatience, any more than pride or levity, in his demeanour; and there was a finer expression of reposing strength, and mild self-possession in his manner, than we ever recollect to have met with in any other person. He had in his character the utmost abhorrence for all sorts of forwardness, parade and pretensions; and, indeed, never failed to put all such impostures out of countenance, by the manly plainness and honest intrepidity of his language and deportment.

In his temper and dispositions he was not only kind and affectionate, but generous, and considerate of the feelings of all around him; and gave the most liberal assistance and encouragement to all young persons who showed any indications of talent, or applied to him for patronage or advice. His health, which was delicate from his youth upwards, seemed to become firmer as he advanced in years; and he preserved, up almost to the last moment of his existence, not only the full command of his extraordinary intellect, but all the alacrity of spirit, and the social gaiety, which had illumined his happiest days. His friends in this part of the country never saw him more full of intellectual vigour and colloquial animation, never more delightful or more instructive, than in his last visit to Scotland in the autumn of 1817. Indeed, it was after that time that he applied himself, with all the ardour of early life, to the invention of a machine for mechanically copying all sorts of sculpture and statuary; and distributed among his friends some of its earliest performances, as the productions of a young artist just entering on his eighty-third year.

All men of learning and science were his cordial friends; and such was the influence of his mild character and perfect fairness and liberality, even upon the pretenders to these accomplishments, that he lived to disarm even envy itself, and died, we verily believe, without a single enemy.

Professor Robison, the most intimate friend of his youth, records that:

When to the superiority of knowledge in his own line, which every man confessed, there was joined the naïve simplicity and candour of his character, it is no wonder that the attachment of his acquaintances was so strong. I have seen something of the world and I am obliged to say that I never saw such another instance of general and cordial attachment to a person whom all acknowledged to be their superior. But this superiority was concealed under the most amiable candour, and liberal allowance of merit to every man. Mr. Watt was the first to ascribe to the ingenuity of a friend things which were very often nothing but his own surmises followed out and embodied by another. I am well entitled to say this, and have often experienced it in my own case.

This potent commander of the elements, this abridger of time and space, this magician, whose cloudy machinery has produced a change in the world, the effects of which, extraordinary as they are, are perhaps only now beginning to be felt—was not only the most profound man of science, the most successful combiner of powers, and combiner of numbers, as adapted to practical purposes—was not only one of the most generally well-informed, but one of the best and kindest of human beings. There he stood, surrounded by the little band of northern *literati*, men not less tenacious, generally speaking, of their own opinions, than the national regiments are supposed to be jealous of the high character they have won upon service. Methinks I yet see and hear what I shall never see or hear again. The alert, kind, benevolent old man had his attention alive to every one's question, his information at every one's command. His talents and fancy overflowed on every subject. One gentleman was a deep philologist, he talked with him on the origin of the alphabet as if he had been coeval with Cadmus; another, a celebrated critic, you would have said the old man had studied political economy and *belles lettres* all his life; of science it is unnecessary to speak, it was his own distinguished walk.

Lord Brougham says:

We have been considering this eminent person as yet only in his public capacity, as a benefactor of mankind by his fertile genius and indomitable perseverance; and the best portraiture of his intellectual character was to be found in the description of his attainments. It is, however, proper to survey him also in private life. He was unexceptionable in all its relations; and as his activity was unmeasured, and his taste anything rather than fastidious, he both was master of every variety of knowledge, and was tolerant of discussion on subjects of very subordinate importance compared with those on which he most excelled. Not only all the sciences from the mathematics and astronomy, down to botany, received his diligent attention, but he was tolerably read in the lighter kinds of literature, delighting in poetry and other works of fiction, full of the stores of ancient literature, and readily giving himself up to the critical disquisitions of commentators, and to discussion on the fancies of etymology. His manners were most attractive from their perfect nature and simplicity. His conversation was rich in the measure which such stores and such easy taste might lead us to expect, and it astonished all listeners with its admirable precision, with the extraordinary memory it displayed, with the distinctness it seemed to have, as if his mind had separate niches for keeping each particular, and with its complete rejection of all worthless and superfluous matter, as if the same mind had some fine machine for acting like a fan, casting off the chaff and the husk. But it had besides a peculiar charm from the pleasure he took in conveying information where he was peculiarly able to give it, and in joining with entire candor whatever discussion happened to arise. Even upon matters on which he was entitled to pronounce with absolute authority, he never laid down the law, but spoke like any other partaker of the conversation. I had the happiness of knowing Mr. Watt for many years, in the intercourse of private life; and I will take upon me to bear a testimony, in which all who had that gratification I am sure will join, that they who only knew his public merit, prodigious as that was, knew but half his worth. Those who were admitted to his society will readily allow that anything more pure, more candid, simpler, more scrupulously loving of justice, than the whole habits of his life and conversation proved him to be, was never known in society.

The descriptions given by Lords Brougham, Jeffrey, the genial Sir Walter, and others, of Watt's universality of knowledge and his charm in discourse recall Canterbury's exordium:

Hear him but reason in divinityAnd, all-admiring, with an inward wish consumed,You would desire the king were made a prelate;Hear him debate of commonwealth affairs,You would say—it hath been all in all his study:List his discourse of war, and you shall hearA fearful battle rendered you in music.Turn him to any cause of policy,The Gordian knot of it he will unlooseFamiliar as his garter; that, when he speaks,The air, a chartered libertine, is still,And the mute wonder lurketh in men's earsTo steal his sweet and honeyed sentences.

If Watt fell somewhat short of this, so no doubt did the king so greatly extolled, and much more so, probably, than the versatile Watt.

Dr. Black, the discoverer of latent heat, upon his death-bed, hears that the Watt patent has been sustained, and is for the time restored again to interest in life. He whispers that he "could not help rejoicing at anything that benefited Jamie Watt."

The Earl of Liverpool, prime minister, stated that Watt was remarkable for

the simplicity of his character, the modesty of his nature, the absence of anything like presumption and ostentation, the unwillingness to obtrude himself, not only upon the great and powerful, but even on those of the scientific world to which he belonged. A more excellent and amiable man in all the relations of life I believe never existed.

There can be no question that we have for our example, in the man Watt, a nature cast in the finest mold, seemingly composed of every creature's best. Transcendent as were his abilities as inventor and discoverer, we are persuaded that our readers will feel that his qualities as a man in all the relations of life were not less so, nor less worthy of record. His supreme abilities we can neither acquire nor emulate. These are individual and ended with him. But his virtues and charms as our fellow-man still shine steadily upon our paths and will shine upon those of our successors for ages to come, we trust not without leading us and them to tread some part of the way toward the acquisition of such qualities as enabled the friend of James Watt to declare his belief that "a more excellent and amiable man in all the relations of life never existed." A nobler tribute was never paid by man to man, yet was it not undeserved.

So passes Jamie Watt, the man, from view—a man who attracted, delighted, impressed, instructed and made lifelong friends of his fellows, to a degree unsurpassed, perhaps unequalled.

"His life was gentle, and the elementsSo mixed in him that Nature might stand upAnd say to all the world, 'This was a man.'"

www.ingramcontent.com/pod-product-compliance
Lightning Source LLC
Chambersburg PA
CBHW081820200326
41597CB00023B/4325